Subterranean Shropshire

Think you might
be interested in
Bridgnorth entries.
Return in due course

love

Daphne.
x x x

Subterranean Shropshire

Steve Powell

TEMPUS

First published 2002

PUBLISHED IN THE UNITED KINGDOM BY:
Tempus Publishing Ltd
The Mill, Brimscombe Port
Stroud, Gloucestershire GL5 2QG
www.tempus-publishing.com

PUBLISHED IN THE UNITED STATES OF AMERICA BY:
Tempus Publishing Inc.
2A Cumberland Street
Charleston, SC 29401
www.tempuspublishing.com

British Library Cataloguing in Publication Data.
A catalogue record for this book is available from the British Library.

ISBN 0 7524 2761 X

Typesetting and origination by Tempus Publishing.
PRINTED AND BOUND IN GREAT BRITAIN.

Contents

Acknowledgements

Special thanks to K. Powell for word processing and typing the document.

Special Note

WARNING: No liability can be accepted for any person reading this book who foolishly tries to enter any of the underground features, some of which are in an extremely dangerous condition and liable to substantial collapse at any time. If you feel the urge to venture underground, you should ask your local library for the address of an experienced caving club. Such a club can introduce you to the sport safely, using the proper equipment and with good training.

The inclusion of underground workings in this book does not infer any right of access. Most of the sites are on private land and anyone wishing to visit should obtain the necessary permission.

Foreword

It was a profound sense of urgency that took over my senses some nine years ago, when I noticed how quickly our hidden heritage of subterranean features was decaying or being destroyed by the forces of nature or the bigotry of vandals. In some cases the underground places had become nothing more than rubbish dumps.

It appeared that although people talked and relished the thought of exploring these features, most having a tale to tell of a visit to a tunnel or old cellar by candle-light or torch, the knowledge and history of them was passing only by word of mouth. Barely anything was being put down on paper other than fragmented pieces. I hope I have now rectified this situation to some degree in presenting to the reader what can only be described as a travel diary of underground Shropshire. I can only express the fun and enjoyment shared by myself, my son and my father, on our 'adventures' as we termed them at the time.

In times past the underground world played a major part in providing housing, food storage and religious sanctity; some people delved in the earth for reasons of pure eccentricity. Luckily, in Shropshire we can account for all of these, including caves formed over many hundreds of years. Rather than hinder any further, I shall say, 'Read on and enjoy'.

Strange Tunnels

As with most other counties Shropshire, or in older times Salop, has its fair share of 'secret subterranean tunnels', most of these tunnels having a basis in the imagination of the storyteller. The Revd I.E. Cartlidge in 1915 made an interesting comment about Wombridge Priory:

> Rumours of the existence of secret passages amount in the locality, but alas, for the love of mystery, the secret passages turn out to be derelict canal tunnels. The absurdity of the idea that the district is honeycombed with secret passages ought to be self-evident.

What became of the vast amount of material removed in the making of the passages? If the mounds of material were visible at either end, what became of the secrecy of the passage? Where were the deeds authorising passages through the lands of other people? What damage threatened the monasteries, which so often passed right of sanctuary? And, one may ask, when did this poverty-stricken priory possess the means to undertake so large a task as the construction of a tunnel to Lilleshall?

The mystery with which the present generation surrounds a monastic ruin is a mystery of its own creating, based upon ignorance of a mode of life which in its own day was well understood as ordinary parish life. (It is interesting to note that in recent years, a roadside collapse exposed an underground tunnel that did indeed turn out to be that of the old Wombridge canal.)

Other monastic buildings of Shropshire also have rumours of secret passages. Wenlock Priory, for instance, is said to link up with Buildwas Abbey. Indeed, in 1867 a heavily laden cart was passing over the court adjoining the Abbot's house when the cart's weight induced a covered archway to give way. On moving the debris from the hole it was discovered that a considerable length of underground passage could be traversed. Georgina Jackson, who wrote *Shropshire Curiosities* in 1883, stated:

> On expressing a desire to see the fabulous subway, by procuring a ladder and candle and on following a guide through a trap door, after going about fifteen feet

*down a vertical brick shaft, we found ourselves in this veritable tunnel. It
appeared about ten feet in height and four feet in width, formed of excellent
squared masonry, and with a semicircular arched covering of well-keyed ashlar.
We followed it for about fifty yards, our further progress being impeded by soil
and the debris of fallen stonework. We noticed one rather puzzling feature, viz.
The existence of a bold jutting stone corbel at about six feet from the floor level,
the only practical purpose that it could have here served, as it appeared to us,
being that of a bracket for holding a lamp to light the passage.*

A later article on Buildwas Abbey states: 'The legendary tunnel is probably no
more than a main drainage adit for the site, turned into legend. Water conduits
of large size may have been needed to take large volumes of rainwater away from
the site due to large roof structures.'

As we can see, we have two types of written account, one romanticised and
one with a practical heading, and it is up to the reader to decide upon their own
conclusions to these matters. Other supposed 'secret passage links' include:

> Shrewsbury Castle to Lyth Hill
> Holy Cross Abbey to Austin Friars House
> Holy Cross Abbey to St Mary's Church
> Holy Cross Abbey to Haughmond Abbey
> Haughmond Abbey to Ercall Hall
> Lilleshall Abbey to Longford Hall

As we can see, the monks of Holycross Abbey were of a very troglodytic nature
and must have spent more time tunnelling than in prayer.

The Shropshire border town of Ludlow is also rumoured to be riddled by a
network of secret tunnels. The Wheatsheaf Inn, one of the older buildings, has
beneath it a veritable warren of chambers with bricked up openings to
underground passages that lead in all directions. One is said to travel from
Ludlow Castle, through the town, beneath the Wheatsheaf Inn, down Lower
Broad Street and finally under the river to Ludford Church.

The Bull Inn, located on Corve Street, is the oldest timber framed house in
the town, dating back to 1343. A story was told to Veronica Thackeray in 1975
by the innkeeper that a small room was discovered when builders were working
on one of the chimney breasts and it was thought the room may have been a
priest hole. It was discovered above the fireplace, together with a flight of eleven
steps leading to the opening of what appeared to be a tunnel going in the
direction of the church of St Lawrence's. It seems very strange that upon such
discoveries there is never any reference to anybody exploring the tunnels or
passages to their full extent when the opportunity arises. Perhaps most people

would rather leave the story open-ended and keep the rumours of endless tunnels in operation, rather than be disappointed in discovering the tunnel or passage ending after a few metres. Again the theory of drainage comes into mind with Ludlow and most of its rumoured passages.

Just outside of Ludlow on the River Terne is found Downton on the Rock and Downton Castle. A hermit's cave or grotto is found to one side of the River Terne at SO 4273. The riverside path runs into the arched grotto, the opening curves round into an abysmal round room excavated out of the living rock. It is of artificial construction, 20-25ft high and has a light shaft on its upper proportions, to spring rays of light into the chamber, the light bringing our attention to a slanting, twisted column, thickening upwards at the end of its spiral. I was told of an identical cave existing on the opposite bank of the river. It is said that the two caves worked in conjunction with each other as a form of custom or policing point to the river in ages past, though another story mentions the cave or grotto being built as part of the garden walks, or perhaps the cave was utilised by the garden walks at a later date.

A solitary dark tunnel breaking out of the limestone hill nearby is that of an old mining tunnel used for extracting limestone for calcination, for usage in fertilisers and building mortar in times past. It is worth noting that these features are on private land and permission to be on this land is needed by the Downton Estate.

Further down river is Downton rock shelter, found at SO 429731. This is noted in the book *Caves of Wales and the Marches* as a place of natural interest.

Moving back towards the centre of Shropshire, we come across the tall house and courthouse in Madeley. There are rumours of escape tunnels leading away from these buildings, but they have not as yet been verified.

Ironbridge has played host to many a strange underground passage; nearly all of these have been attributed to various mines for mineral extraction. However, an interesting place is The Lodge near Lincoln Hill, which was built in 1530 from limestone blocks. It has walls nearly 2ft thick and it is rumoured that two priest holes were added to its construction, while in the cellar there is a tunnel entrance, that is now bricked up.

Secret passages were known to exist at Ketley and Burrows Farm at the turn of the last century. In 1983 Mrs June Colley from Kingswinford said that it was then a crumbling building and that the farmhouse had a cellar. Her husband said there were passages inside and it was rumoured that these led to the old summerhouse and another to beneath the Bell Inn in the High Street. He never went along them as he was too scared, being only a child at the time, but he remembered the openings.

Near Alveley is a stately home known as Coton Hall. This is reputed to have been built on one of England's oldest estates, dating back to the Roman era.

The hall is built on a solid bedrock foundation that is reputedly honeycombed with tunnels. It is said that one of the wine vaults has been blocked off to hide one of the secret tunnels. There appears to be no mention of why the tunnels exist or their destination.

Bordering upon Wenlock Edge is Shipton Hall. Local history writer Michael Raven mentions in his *Shropshire Gazette* that 'there is a tunnel from Shipton that emerged in front of the fireplace at nearby Larden Hall, it is now only passable for a few hundred yards.' (Larden Hall is now demolished.) The earliest reference to any sort of passageway was in 1899 by H.T. Timmins, an avid traveller around Shropshire. On one of his journeys passing through the village he noted that, 'As we start back to Shipton we noticed a curious sort of grotto or cavity in the limestone rock.'

A later description written in 1915 by H.E. Forest in *Old Houses of Wenlock* states that Shipton Manor belonged to Wenlock Priory up to the time of the dissolution. Forest also adds that '[a] tunnel near the hall, now half full of water has in modern times been described as an underground passage to Wenlock Priory.'

Shropshire Caving and Mining Club explored a passage in 1978 which was found in a limestone cutting below Shipton and is possibly that mentioned by Timmins and Forest. The passage travelled 55 yards to a collapse and this was dug through to gain a further 40 yards of tunnel, which was still ongoing. Exploration had to be curtailed due to narrowing air space, that was now down to 10cm, and also due to the consistency of thick, silty water flooding most of the tunnel, making progress unpalatable. The landowner, Mr Bishop, remembered as a youth trying to take a small boat up the flooded tunnel, but as the ceiling height became smaller, he was forced to abandon exploration. Also, the strange phenomenon about this tunnel was that it slopes gradually downhill into the earth, explaining why the water eventually backs up to the ceiling. It was thought at first that the tunnel was a mining adit, but usually mine adits run the ground water out towards the entrance, so as to leave the tunnel self-draining. At this present time its function are unknown.

Alan Taylor of Shropshire Caving and Mining Club mentioned that Mr Bishop told him that as a young lad he had lived at Larden Hall (c.1905–1920) and he remembered a cellar/hatchway being present in the kitchen floor that led to a passageway. Mr Bishop later moved into Shipton Hall and remained there until his death in 1996/97. So there we have it, references to entrances at both Shipton and Larden, the connection between the two unfortunately not at this time proven.

Another building of antiquity with rumoured underground features is Meeson Hall at Great Bolas, which has been dated back to 1640. Back in 1995 I went along to Meeson with an investigatory group from Shropshire Caving

Wild Edric riding out from his cave.

and Mining Club to try to conclude its secret tunnels, which are presumed to travel to many local buildings, some as far away as two miles. A dowser was also present on this occasion, spending two hours hoping to uncover anomalies in the surrounding grounds. Only one tunnel was found; this begins from a cellar cut out of the sandstone bedrock beneath the house. Various drainage channels were found in the floor of this chamber, an outlet (blocked by a large paving slab) connects into the rock-cut tunnel of man-sized proportions. This then travels for approximately 200m to a collapsed zone in nearby woods. An entry grid was found connected to the tunnel part way along it s length and surface depressions in the wood lead to a small stone-built chamber. Dave Adams, founder member of Shropshire Caving Club, states how he thinks the system was that termed a wet cellar. In the seventeenth century many great houses brewed their own beer. In order to keep it cool a spring would be tapped into the cellar and continuously run off into a drain, this would have the desired effect of consistent cool temperatures suitable for the beer and also any dairy produce. It is possible that a second cellar existed in the property that is now lost.

There are two underground passages existing in Shrewsbury, unfortunately neither of them are accessible, but they are both mentioned in *Shropshire Notes and Queries* dated January 1885, which states:

One passage leading from a house belonging to a Mr Beacall, on the old wall on the north side of Castle Street (formerly Old High Pavement). The other from 'Vaughans Mansions', in which the town's museum was later situated. The tunnel travelled south under College Hill towards the former site of St Blaise Chapel in Murivence. There is no mention of them being explored or their dimensions.

Folklore tales of secret tunnels abound near to the Shropshire borders. A couple of these are mentioned by the author Ida Gandy:

The rock of Woolbury above Clun – that's a grand great quarry, so big you could put the village inside it, they say all the stone for Clun Castle came from there. The big rock at the far end is called the Pulpit Rock, there's a deep cleft nearby and when you drop a stone into it, you can hear it going down and down. They say there's an underground passage there that leads all the way to the castle.

There are no facts at present that confirm this tale, but near to Clun is a place known as Skyborry, and an old folktale states that years ago a man's skull and a great key were found in the cellar, but none could open the door leading out of it. The cellar, they say, is linked to an underground passage that goes all the way to Craigdun Rocks. Above Knucklas, to the side of the river, an old man who spun the tale said that, 'There's a hole up there, you can see it now, I've put my stick in and there don't seem no bottom to it.' A few years later the old man asked Ida if she had yet been up to find the great hole in the rocks above Skyborry. One can sense that she never did, as confirmation to its existence was not given in her book *An Idler on the Shropshire Borders* printed in 1970.

Another tale is told of secret vaults holding a treasure trove in the rock outcrop, beneath Stokesay Castle. Various archaeological digs have been completed at Stokesay Castle, and I feel sure that if an underground passage and chamber existed they would have been discovered.

The next tale is of a mythological figure called Wild Edric and, like most old tales, its telling changes into various versions. Georgina Jackson, in 1883, had Wild Edric living in old mine workings beneath the Stiperstones, while a more modern version states how Wild Edric, the Anglo Saxon lord, led an uprising against the Normans in 1069-70 and he now lives buried in a great cave, awaiting the day when an English king once again sits on the English throne. He is said to ride out with his ghostly army from his cave in the Stiperstones if war threatens. Georgina Jackson also mentioned Wild Edric riding out for similar reasons. She also gives us a full name of Edric Silvaticus.

This folktale is in very similar veins to the ones about King Arthur and his knights who also lie in deep enchantment awaiting a final battle.

14

Tong Castle in East Shropshire, built in phases from the twelfth to the nineteenth centuries, was host to many underground structures and tunnels, some cut from the living rock and all interconnected. Various cellars, larders and some of the passageways were joined together, and it is said that in the eighteenth century the servants could move and live completely underground, out of site of the castle, owner and guests. However, there was very little ventilation to the tunnels and cellars, so life would have been most uncomfortable. There were two large cellars, one reputedly big enough to hold 360 dozen bottles of wine, the cellars being cut deep into the sandstone bedrock. A spectacular vaulted chamber with arched ribs for roof supports was found to the south-east corners of the castle grounds and this may have been used in conjunction with a watchtower. Unfortunately, during 1954, the whole of the castle was demolished for safety reasons. However, the tunnel systems were excavated and reported during archaeological digs undertaken by Alan Wharton during the 1970s. The reasons for Alan's survey was that the M54 motorway was scheduled to cut through the site, losing a lot of its archaeology. Most of the area left after the motorway was built is now much overgrown and backfilled.

There is presumed to be a secret tunnel from the Friars to the Fosters Hotel in Bridgnorth and another secret tunnel from Upton Cressett Church to the nearby manor house. The Bridgnorth area is a myriad of caves and tunnels and is discussed in a later chapter.

An interesting story came to light in 1990 about Heath House, a twenty-five room Queen Anne mansion that had been occupied by nearly blind Simon Dale, who was unfortunately murdered. According to the weekend *Guardian*, Dale was an eccentric who was digging the grounds of Heath House for an Armenian City or Camelot. His main hobby, however, was exploring the tunnels found beneath the house and excavating his kitchen floor. His near-blindness necessitated enlisting the help of a friend, who said, 'You might say he had a fetish about those tunnels, but just because you're interested in tunnels it doesn't make you mad, does it? If I had a load of tunnels in my cellars and nothing better to do, I'd poke around in them.'

There is a tale in Alveley that a subterranean passage travelled from the cellars of what is now the church cottages to the chancel of the church. The monks of Coton College are presumed to have stabled their horses in the cellars and then travelled on to church via the underground passway.

Upton Cressett Hall, constructed in 1580, is said to have had a subterranean passage connecting to Holgate Castle in Corve Dale. The reality of this connection must be very slim, as the distance between the two properties is at least six miles. The well-respected H.T. Timmins said in 1899 that this particular enterprise must be accepted 'cum grano salis'.

Plowden Hall, built in 1557 by Edmund Plowden, is said to have been 'honeycombed' from cellar to roof with secret chambers, passages and hidey holes. A subterranean tunnel is said to have led out to a lonely spot in the woods known as 'Lady's Chair'. In the house itself is a 2ft wide shute, circular in shape, that runs between the thick set of walls from top to bottom of the house. A person could effect escape from it with the aid of a rope. It is said of Plowdon that the builders had in mind the Spanish proverb: 'The rat that has one hole is easily caught'.

Gate Acre is another place associated with secret chambers and passways. The Earl of Derby is reputed to have hidden in the grounds after the Battle of Worcester in 1651.

Llanymynech Ogof

Although technically over the border of Shropshire by the smallest of margins, Llanymynech Ogof is found at grid reference SJ265219. It is included because it is a cave that has been of great interest to the people of Shropshire for many generations and is in modern times the most easily accessed and used cave in the area. Having entered the cave's main chamber, it splits rapidly into a labyrinth of interconnecting chambers and passages. Most of the passages, being very low, flat-out crawls, have been given names to testify to this, such as Agony Crawl. There is one very low passage that leads to an extension known as the water series. Here the caver can enjoy a low belly crawl with only minimal air space, the good side being that the passages and chambers do get larger once through this constriction. There are in places some static formations, small straws and isolated rim pools. There are large cascades of flowstone in a place called 'Shaft Chamber'. The most interesting section of the Ogof is a section known as the Figure of Eight Squeeze, which winds through partly flooded passages. Again formations are in evidence in this section, which is harder getting out of than getting in.

The cave passages are mainly old mine workings, dating back to the Bronze Age and Roman period and have great historical background. In times past, coins and artefacts have been found, confirming this. The passages have been dug out for the mineral copper which has been found as malachite and azurite. During my own explorations, a small bell chamber past the water series had a beautiful section of deep blue azurite embedded in its sidewall. I have heard from a friend that this has now been recently hacked out by some uncaring caver. The minerals have been mixed in with carboniferous limestones, which are usually host to natural caves.

Some solutional cavities are in evidence, in fact some of the mine passages may well have been widened from existing natural cavities, to exploit these minerals.

Local geologists have stated that the large amphitheatre that lies below the Ogof was formed by the collapse of an ancient cave system. Early explorers to the cave around 1890 to 1920 state how a stream used to sink into the ground on this plateau, but since the formation of the golf course many of these type of features and entrances have been lost.

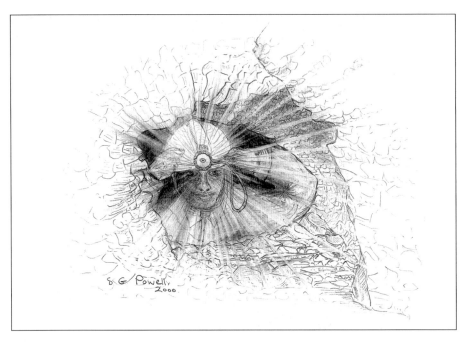

Caver in the Ogof.

Not far away from the 'Ogof' entrance is an old mine adit from the nineteenth century, this too, leads to interesting passages. The adit is part flooded in its entrance and various side passages lead away at ninety degrees. One of these travels a great distance into very unstable places where sand slopes are climbed into upper cavities. On one occasion, I climbed under a boulder ruckle at what was presumed to be the terminus shown on a cave survey, and an extension of a further 150m of mine tunnel was followed through unstable ground to a choked end. There were signs that other cavers had reached this point also.

There is an upper series, which is accessed via an internal shaft at the adit end (termed a whinze), having climbed the 20ft shaft, passages wind out in two directions following the mineral veins. These passages in the upper series are low and constricted and once again are presumed to have been excavated by people many centuries ago. Bats frequent this upper series during the wintertime and they should not be disturbed.

The 'Ogof' is surrounded by much folklore. An early explorer to the cave, J. Dovaston, at the beginning of the nineteenth century, commented, 'Superstition, ever prone to people in darkness with the progeny of imagination, has assigned inhabitants here, such as knockers, goblins and ghosts and the surrounding peasantry ever with inflexible credulity that the aerial harmonies of fairies are frequently heard in the deep recesses. Tradition says this labyrinth is communicated by subterraneous paths to Carreghova Castle and some persons aver that they have gone so far to hear the Rivers Vyrnwy and Tanat rolling over their heads and that it leads down to fairyland.' A later transcript taken from *Brayleys Graphic and Historical Illustrator* of 1878 relates how the cave is supposed to extend for an endless distance underground and was of so fearful a nature that it was reported that any persons venturing within five paces of its mouth would infallibly be lost. It is said that animals also had a great fear of the cave and that a fox in full flight from a pack of hunting hounds would turn and run into his canine enemies rather than encounter the cave entrance. The hounds however would not touch the fox, as it was now tainted for imbibing the Ogof's powers.

Climbing the Whinze.

Several human beings were believed to have been lost within its ponderous and marble jaws. One of these was an old minstrel, who fell victim to a rash bet on the subject. He danced towards the cave till he came within the limits of its charmed circle, when he was suddenly seized by an invisible power and hurried away for ever from the gaze of man.

It is later inferred that the cave was the abode of a fairy that became the wife of a King Alaric, whose palace is said to lie at the bottom of Llyncly's Pool.

The same journal published a further story in 1896 that stated:

The Roman cavern in Llanymynach Hill, called Ogo, has been long noted as the residence of a class of the fairy tribe, of which the villagers relate many surprising and mischievous tricks. They have listened at the mouth of the cave and have sometimes even heard them in conversation, but always in such low whispers that their words have never been distinguishable. The stream that runs through it is celebrated as being the place in which they are heard to wash their clothes.

A most famous tale which also appeared in 1896 refers to Ned Pugh. Ned asserted that he could walk underground from the Ogof to the Lion Inn at Llanymynech village. He was not believed, but when he then made a wager that he would, on the following Sunday, play a tune at the usual time that the choir sang and that he would be heard by all the congregation in church, his boasting challenge was taken up. On the following Sunday Ned, carrying with him his harp, went to the entrance of the Ogof on the hill and disappeared within. As the time came on for the choir to sing, everyone was intently listening for the sound of the harp and sure enough, out of the earth, proceeded its sounds. The people distinctly heard a tune, which the singers took up, and when they had finished the harpist too ceased, the poor man though never emerged out of the Ogof. The tune in consequence was called *Farewell, Ned Pugh.*

Oreton/Farlow Cave

Shropshire is not a place usually associated with natural caves due to its irregular geological strata and geological positioning. Nevertheless, unknown to many, there are small numbers of explorable cave passages to be found. One of these sites is on the north-eastern side of the Clee Hills and found in the carboniferous limestones. The Shropshire Caving and Mining Club began investigations in the area as far back as 1964 in which it was noted that the resurgence of an underground stream was found to be running strongly after heavy rain and needed closer attention. The resurgence at Oreton was again looked at in December of 1965, but not followed any further. This was left to a caver, Duncan Glasfurd, from nearby Ludlow. Duncan became quite a celebrity in the area (as we shall see later) with his exploration and investigation techniques. The cave system begins at a place known as Silvington Woods, near to the Rylands, SO 6287966, where a line of sink holes can be found travelling in the direction of Farlow. When Birmingham Enterprise Club investigated these in 1970, they were found to be intermittently open but had no prospect for underground exploration. The main sink is found at Farlow and is termed Foxholes Swallet, SO 641803. Many attempts have been made to gain access to underground passages at this point. George Price of the Birmingham Enterprise Club stated in 1970 that:

> The arenaceous oolitic limestone is structurally weak in the vicinity and the stream passage is blocked by a boulder ruckle and flood debris after a few yards. The dry way shaft sunk by the late Duncan Glasfurd further downstream was examined, but substantial supports would be required to render the shaft safe for further excavation. Anyone hoping to strike the stream way by this route is warned to take extreme care, the reversal of the final flat out crawl is difficult and the rock strata unstable.

Indeed by 1980 the Shropshire Caving and Mining Club mentioned that the Foxholes cave system was now un-enterable, and during 1999 the land owner filled in the dry way shaft for the safety of his children. The underground stream travels from Farlow to Oreton, a distance of 1.6km with a fall in ground level

Glasfurd's Passage.

of approximately 45m. Further sink holes are found in the old limestone quarries between Lower Truckhill and Oreton bank and George Price states how they close down to tight fissures with boulder ruckles.

After being unsuccessful with the dry way shaft, Duncan Glasfurd later found a way into the underground stream way near its resurgence by forging an incline tunnel to intersect the natural passage. This was found to travel 45m towards Farlow before coming to an abrupt halt at a water-filled pool or sump. It is said that Duncan attempted to free dive the sump to continue exploration but was unsuccessful in finding a way through. The Birmingham Enterprise also dived the sump and commented that at this present time it is very hazardous, due to extremely poor visibility due to movement of silt deposits.

Duncan, after this discovery, took many local people into the stream passage and was highly respected. Unfortunately Duncan lost his life in Yorkshire exploring the cave systems there. The stream way at Oreton is now known as Glasfurd's Passage. Friends Alan Robinson, Colin Bradford and myself, made a visit to the cave on 18 April 1999. Alan had visited the site before and led the way to the entrance of the stream resurgence at SO 657803. This is impassable as the water is released through a fissure between separated strata beds. However, an entrance is conveniently situated beneath a tree whose roots lead into the main stream way a few metres upstream of the resurgence (SO 656804).

The entrance is quite constricted and is similar to the S-bend of a sewer pipe. The floor drops 10-15ft sharply downhill to enter a passageway through a boulder-strewn floor. Colin was unfortunate in being too large across the chest to get through the entrance, so Alan and myself were left to exploration. This entrance passage is dammed off at its junction with the stream, thus diverting all of the water into a sumping passage to the resurgence and keeping the explorer temporarily dry. We climbed over this boulder and clay dam and set off upstream, crossing two further boulder dams or rock piles to enter a small chamber approximately 30m further on. There the stream disappears beneath a pool which, on this occasion, was approximately 4ft deep, although evidence in the passage suggests its total flooding on occasions. A few photographs were taken before exits were made to surface. It was noted that an enormous amount of clay and silt had been deposited throughout the cave system. There also appeared to be a far greater flow of water in the stream way than what sinks at the base of the cliff at the Fox Holes, Swallet, further up the valley, perhaps indicating a further inlet.

The Birmingham Enterprise Club attempted to bypass the sump by setting off explosive charges in the hope of widening the terminal passage, but was unsuccessful. At present, possible extension to the cave system looks extremely doubtful.

Sir Roderick Murchison, the famous geologist, gave an insight into the area in 1839. 'The Oreton limestone is cut by many transverse faults, at which the sides are thrown up at different angles of inclination. The Farlow Brook runs through the fissure produced by one of these faults.' A further article appears in the geologist's summary of Morris and Roberts in 1862 entitled 'Carboniferous Limestone of Oreton and Farlow'.

We are indebted to the Rev. J. Williams of Farlow, for some valuable information relating to a recent exposure in one of the deepest of the Oreton quarries, of the subterranean stream, which has long been known as flowing parallel with the axis of the ridge. This Mole River loses itself in a hollow called the Foxholes, at the western extremity of the limestone and taking a N.N.E course reappears at the distance of a mile. Two of the quarry men, who had struck upon it at the depth of about 50 feet from the surface, described it as a constant stream, occasionally greatly swollen by floods. An interesting account of an accidental stoppage at its inlet during one of the great floods of last year was furnished to us by Mr Williams. He stated from his own observation that two and a half acres of the hollow were covered to a depth of 15 feet by the damming up of its accumulation

Entrance to Farlow Cave.

of water through its underground passage. From the data supplied by the careful observations of Mr Williams, the lake thus formed must have contained 1,635,000 cubic feet of water and the ratio of its subsidence was not less than 34,000 cubic feet of water per hour. It appears from this that the fissure though which the stream flows is of no insignificant dimensions.

There are possibilities of a further natural cave system a few miles away at Brown Clee, where a stream runs into a very large sinkhole that diverts the water underground besides the road and car parking. Water can be heard gurgling and dropping steeply into a cavity that is, at present, part blocked up by silt and flood debris. This sink is found in the Abdon limestone at SO 608872 and requires further investigation.

Aldon Fissure and Ippiken's Rock

Both of these features were found in the much earlier Silurian limestones. The first, Aldon Fissure, was a vertical joint irregularly opened out by the uplifting of the bedding plains in some bygone epoch. The importance of the fissures was not that they might have had underground access, but that the joints were filled with a calcareo-argillacous cement in which the remains of ancient animals were found. It was stated in 1839 by Roderick Murchison that stags' horns and bones of great size were found. Unfortunately, at the time, the area was being worked for stone for lime burning and most of the bones first discovered were obliterated.

The fissures and finds were discovered and reported by Mr Duppa Lloyd and Dr Lloyd of Ludlow, who recorded the remains of rhino, red deer and a perfect tooth of a hyena and other unidentified species. The famous geologist Roderick Murchison stated:

> I would also further observe that there are also true caverns in which the remains of extinct animals may probably here after be found, such as Ippiken's Rock on Wenlock Edge. Another remarkable ball stone (not yet consumed by the lime burners), it constituted a boss, the summit of which is about 400-500ft above the alley, occupied by the Wenlock shale. The rock itself has a vertical face of 50 or 60ft, which is partially fissured and presents an entrance into a small cavern. A person of the name of Ippiken, tradition says, formerly inhabited this cavern. The cave has been recently examined, at my request, by my friend Mr R.W. Evans, to ascertain if it contains any animal remains. Mr Evans, of Kingsland, found in the cave, layers of alluvial deposits, there were no organic remains. He was, however, unprovided with sufficient assistance to make a thorough examination. It was hoped at this point that the Ludlow Natural History Society would complete the survey and research.

G. Jackson in *Shropshire Folklore* briefly mentions the cave in 1883:

Between Presthope and Lutwyche Hall, the cliff is connected with a mythical hero, Ippiken, a famous robber knight of days gone by, who inhabited a cave at the base of this crag, concealed among the trees and brushwood. The mouth of the cave is still open.

H.T. Timmins told a more interesting tale in 1899, during his rambles around Shropshire:

Setting our faces towards Wenlock, we now follow a high-lying Ridgeway Road, commanding fine views in the direction of the west. We strike into one of the numerous footpaths and make for a sort of cave, or rather cranny, high up in the limestone rocks of the edge.

Amidst tumbled boulders and brushwood, this is Ippiken's Rock, the haunt of a robber knight of that ilk, whose deeds were famous in days of yore throughout all the countryside. Here, as the story goes, Ippiken was wont to foregather with his merry men all, issuing forth and levying blackmail on passing travellers and hiding the stolen treasure in the rocky fastness, where the print of the knight's gold chain it is said, may still be seen. Strange lights twinkling like will-o'-the-wisps at dead of night, struck terror into the heart of the country folk as they gazed in fear and trembling from the rustic homestall, while Ippiken and his crew held high revelry in their unapproachable eyrie. Eventually Ippiken himself was slain and his band dispersed, so that they troubled the king's peace no more, but if tales be true, Ippiken's ghost still revisits for glimpses of the moon and may be summoned from the vast deep by anyone who cares to stand atop of the cliff at midnight and cry three times 'Ippiken! Ippiken! Ippiken! Keep away with your long chin.'

As with most folklore tales the story as a whole remains the same, but with slight variations as time passes from generation to generation, as can be seen from the next version told by the rabbit catcher William Southall, a well-known character of the Shropshire district, to the author William Byford Jones in 1937:

Down there, said the old man, You see a rock, it's Ippiken's Rock, old Ippiken was a bandit, a real bold bad lad, in these days e'd be giving the coppers a hot time chasin' 'im away from banks. 'E used to live in a cave there, and 'e robbed everybody with money round 'ere in the old days and in that cave which was his stronghold, 'e kept a sight of cash. One day there was a thunderstorm and old Ippiken and his robbers were in the cave when the lightnin' struck it and smashed it in, they were all buried wi' the money an' they died together as robbers should. And his ghost? Ah if any man goes down there,' said the rabbit catcher, 'And cries out Old Ippiken? Old Ippiken? the ghost of the robber will come and he'll slash out with a chain.

The cave in question must have collapsed between 1883 and 1937 and this had been drawn into the tale. Ippiken was presumed to have existed during the fifteenth century and obviously rumours of buried treasure were unfounded, as proved by the archaeological digs around 1839. The site, at grid reference SO569965, was looked over by myself in 1999. Sadly not even signs of a reasonable fissure were found along the cliff and it's a pity that nobody gave a description of its interior.

Alderbury Fissure Cave

On 2 March 1997 several members of Shropshire Caving and Mining Club looked at a fissure cave in red sandstone near Coed Way. It is found in a cutting on the south side of the B4393 at SJ 354145.

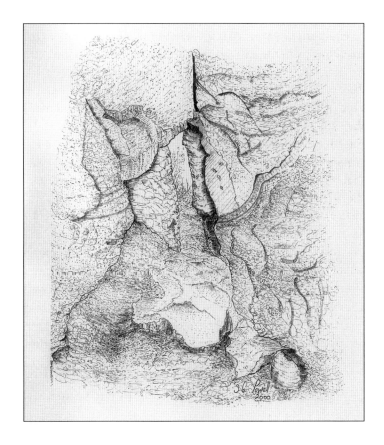

*View inside
Alderbury.*

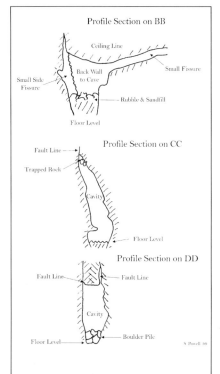

Sections on Alderbury.

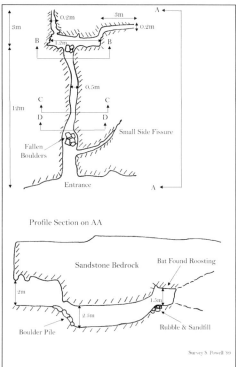

Plan of Alderbury Cave.

Nick Southwick explored the passage, which is walking height at the entrance and drops to a crawl, but then opens up on the other side. It runs for about 15m to where it ends with two small fissures going up into the roof – a hibernating lesser horseshoe bat was once seen at the end.

At the later date of 27 December 1999, I decided to have a look at the cave for myself. The cave entrance was easily found and easily accessible. The cave is much as described by Nick Southwick. I shall further expand on this in that the fissure cave is of natural origins. The passage is tall but narrow in width; a double fault line is located in the ceiling just inside the entrance, which meanders into a single vertical fault line and I conjecture that it travels through the bedrock to surface approximately 5-7m above. A solitary bat was again seen hibernating in the small chamber to the rear of the cave, but there did not appear to be any other noticeable fissure or caves in the vicinity.

Ceriog Cave

This is found in the carboniferous limestone belt near Chirk Castle. The entrance is situated 15-20ft above the River Ceriog at SJ 265376. Suitable parking was found at the old disused lime kilns on the B4500; the kilns are situated directly above the caves.

The entrance to Ceriog Cave is very picturesque, being surrounded by ferns and trees, and makes a perfect artists' canvas. This large-looking entrance soon peters down to a walk-in sized passage which turns in a dog leg fashion opening up into an unexpected large chamber. There was a profusion of bats found here on our visit in May 2000. It is probable that they stay in and around the cave all year round, so it is best not to disturb them unnecessarily. The way forward, once again, quickly peters down to a fissure-type rift, which is traversed, occasionally opening out to very small chambers. The cave passage eventually splits into two routes, a low level water and mud filled tube and a higher, very tight rift passage, the explorer having to contort into all sorts of gymnastic positions in order to squeeze between boulders blocking the way through the rift. The two routes connect up again further on into the hill, where the cave quickly spreads its veins into a small labyrinth of interconnecting tubes of thick glutinous mud. Recent heavy rain showed a very strong connection to the surface by the deposition of large leaves and foliage still freshly green. My son Lee found the cave very interesting and sporting, but it is certainly not for heavy built persons in its inner sanctity.

The cave does not appear to have had any historical anecdotes connected to it. The earliest written transaction dates to 1905 when an article was written by an anonymous E.A.P for the *Liverpool Daily Post and Mercury*. The cave, it was reputed, extended over six miles in the direction of Oswestry. However, the farthest point reached was a little over 500ft. Local adventurers had got in nearly 100ft and had left plenty of evidence in the way of match stubs and candle grease, and also marks of crawling and scrambling. E.A.P and a friend penetrated the cave to a small chamber with four exits, each of which they explored, every branch led back to other points of divergence or to small tunnels or pipes through which water flowed in rainy weather. There are, in parts, stalagmite curtains and small stalactites.

E.A. Baker and H.E. Balch, in the 1907 publication of *Nether World of Mendip* mentioned the cave next. This might seem strange considering Shropshire and Mendip are many miles apart and totally separate caving areas.

A detailed description of Ceriog was made by North Wales caver Peter Wild, in an edition of *British Caver*, Vol.9, 1942. Further written transcripts were made a few years later by P. Wallis in 1950 (*British Caver* Vol.20 and *Belfry Bulletin* No.29) and a more recent account by Alan Ashwell in *South Wales Caving Club Newsletter* of August 1959. All of the descriptions mention the difficulties encountered, such as: 'At any rate, we were very tired when we emerged and after only two hours of caving.' On reflection, however, I can recommend a trip.

A final report was made by Shepton Mallet Caving Club members F.J. Davies and B.M. Ellis in 1960 in an offprint entitled *Caving in North Wales*, whose cave survey can be seen on the opposite page.

The Shropshire border area also, at one time, boasted an upper Ceriog Cave, which was situated a mile to the south-west in an old limestone quarry at a higher elevation some 1,000ft above that of Ceriog Cave.

Upper Ceriog was first mentioned by Balch and Baker in 1907, who stated:

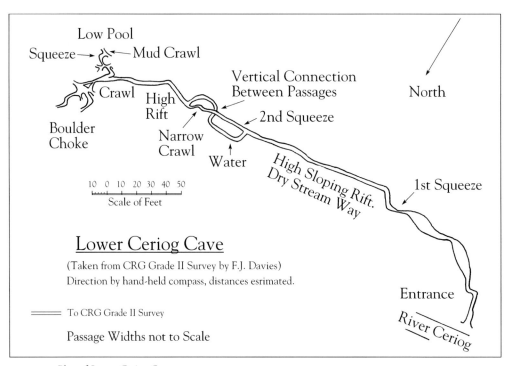

Plan of Lower Ceriog Cave.

At the top of the hill, in a disused limestone quarry, there were traditions of a cave opening that had been covered by a landslip for thirty years. A man was set to work in digging it out, and a small fissure was disclosed, the old channel of a tributary leading to the middle of a cave running N.N.E and S.S.W. The total length was 172ft. The water, apparently at the top of the left passage, ran away into a low bedding cave to the right. The floor is wet clay at present, but there are traces of large stalagmites including one handsome beehive and the roof is covered with beautiful white and amber stalactites.

The cave entrance was eventually lost until 1961 when Dr Gordon T. Warwick of Birmingham University and geologist to the Cave Research Group relocated the site of the entrance, which was again buried by landslip.

This was dug out by a team from the South Wales Caving Club, led by Alan Ashwell, who also mentioned a tale told by a local, reputing the cave to go right through to Bron Y Garth. Apparently one summer, a local farmer had many gallons of milk go sour and he decided to dispose of it down this cave, and for several days following this a certain stream in Bron Y Garth was coloured white.

The South Wales Caving Club did not succeed in their dig, but handed it on to Shropshire Mining Club, who broke through after three hours work. M. James led a party through a tight squeeze down a steep slope covered in thick red mud and this gave access to the side of a water-worn passage some 8ft in width and varying in height from 3-5ft. The ceiling was covered with a mass of small stalactites averaging 2in in length, a clean but debris covered floor sported many large stalagmite bosses and several small helictites were also noticed. It was stated that the cave held possibilities for extension, but the landowner was very hostile to any exploration and has since buried the entrance in many feet of rubbish, leaving the entrance still un-enterable nearly thirty years on.

Canal and Mining Tunnels

The canals are once again a flourishing factor around the country as well as in Shropshire and, inevitably, the canal network has underground features. However, I shall not choose the systems that are still in use, but interesting ones that have been long abandoned.

The first canal feature and associated rock cut chambers we shall discuss are those found at Erdington near Bridgnorth. In fact, so obsolete and obscure are these features that during my reconnaissance and fieldwork, people of the locality were amazed that I had found out about their existence. And after a much-puzzled look by one particular local, he sent me on my way with references to two canal tunnel portals, stating that the tunnel had now collapsed. The underground canal was formerly constructed to supply water from Erdington upper forge to the lower forge under the base of the sandstone cliffs near to the River Severn. The use of sluices and an overflow passage cut into the back of the sandstone cliff gave the water power to drive machinery. The River Severn was unable to do this due to fluctuating water levels, although goods were brought through the tunnel to the upper forge, thus giving the tunnel a dual function. The sandstone escarpment was also used to good advantage and various large caverns were excavated out of the solid bedrock and interconnected for the purposes of workshops and stores. Many of these caves and the portal of the canal can still be seen, but are used and owned by nearby private residences, in fact, a humble garden shed can be seen sheltering inside one large cavernous entrance, which quite overpowers its occupant.

The forge and tunnel were first looked at by W. Byford Jones when he wrote *Both sides of the Severn* in 1932. The tunnel was constructed in the eighteenth century and its water supply came from the Morbrook. The brook had been dammed off to form a weir type structure, in which a large iron pipe from the dam to the upper forge wharfage conveyed water. Water still runs out into the wharf to this present day. The canal then runs underground for approximately half a mile in a south-easterly direction, before appearing to daylight behind a brick-built dam that abruptly suspends the canal about 20ft up in the cliff face.

Cave entrances and canal tunnel, Erdington.

The water overflows romantically, cascading down through a brick-arched weir to the outside world.

A roomy wharfage has been hollowed out of the rock on the northern bank of the lower forge exit for loading and unloading goods, including wine, nails and iron bars that were transported along the tunnel. A winch is in evidence at the upper forge, indicating that boats were hauled through from the lower forge to the upper forge and then returned under the natural flow of the water current. The upper forge, it is said, became obsolete in 1790, whilst the lower forge became obsolete *c.*1889.

The tunnel became of interest to the Shropshire Caving Club in 1965 and, after various arrangements, a trip was arranged for February 1966 to explore its entirety. The trip also included an *Express and Star* newspaper reporter. The tunnel is of small bore, being only about 5ft in diameter, and the water takes up half of this, leaving approximately 2ft of air space. The club members were told of the tunnel's collapsed state, which was presumed to be silted up. They were also told that all other previous attempts at traversing its length had failed miserably. Undeterred (as cavers usually are), the party set sail into the unknown. The first boat was manned by the founder member of the caving club, D. Adams, who was joined by K. Locke and the photographer. A second boat followed, manned by C. Lears. Progress was made by using hands and feet against the ceiling (or 'legging').

The roof lowered to 18in in places and the bottom, when tested, revealed 12in of slimy silt. About a third of the way through there was a chamber in which one could stand up. After some time daylight was sighted, progress however became difficult when silt left little more than a few inches of water. The boat containing C. Lears turned back at this point. However, after sheer brute strength, the lead boat forced its way to the opposite entrance, which was little more than 15in high. The journey had taken fifty minutes. The boat was dug out and turned round and a fresh crew then completed the return journey in a much shorter fifteen minutes.

I should imagine that the tunnel is still open for its full length even today, as a good flow of water still makes its way through to the lower forge portal.

There is a large tunnel at Knowle Sands near Erdington that cuts through the sandstone outcrop. The tunnel forms a route for the Severn Valley Railway. During the severe rainstorms of November 2000 the River Severn rose to unprecedented levels and made use of the tunnel as a through drain, washing away the track bed and buckling rails in the process.

Also at Knowle Sands is a cave cottage called The Whitehouse. A family friend, Mr Dave Richards, explained how, back in 1957, he lodged at the cave cottage for a full week. He was a young fifteen-year-old at the time and had been invited over by his friend John Grainger, whose family had just moved into the property.

The cave cottage front was of red brick, coated with whitewash, hence the name The Whitehouse. Dave remembered the dwelling as being similar in appearance to the rock houses of Kinver. It was located in a coppice on a sandstone outcrop. The house was in view of the River Severn and reached by a pathway that started from an open space near The Swan public house, situated on the Bridgnorth to Highly Road, the distance along the path being some 400m.

The accommodation consisted of a living room with fireplace and provision for cooking and heating water. There were two bedroom enclosures, the front bedroom having a river view, with the second bedroom at the rear excavated further into the bedrock.

The furniture was normal for the period, being mostly pre-war with a few utility items. The water supply came from a standpipe some 100m away and the water had to be carried in galvanized buckets to the house.

Toilet facilities were typically rural, being basically an outhouse (termed the Thunder Box) that had a wooden seat with a hole in the middle that was placed onto a bucket or receptacle that could be emptied when necessary.

Dave also mentioned that his father had regularly pitched an army bell tent every year in the locality from 1947 to the late 1960s. This was used as a weekend retreat for fishing excursions. He knew the area well and visited the caves at Erdington on odd occasions.

The Berwick canal tunnel, another forgotten place, was constructed in 1797 and is named after Lord Berwick, whose lands it travelled through.

The canal consortium, at that time, originally intended to dig out a large cutting to accommodate the new canal, but Lord Berwick was thoroughly opposed to the idea, as it would have intruded into his game reserves. However, he did sanction them rights to put the canal beneath his grounds and so the tunnel of three-quarters of a mile long, brick-lined throughout, was constructed. The tunnel bore is 10ft in diameter, and as with most early canals, there is no towpath. The boats were propelled through the tunnel by legging. The horses that usually towed the boats had to travel by road and meet them at the opposite end.

The tunnel has imperfections, the most serious being that it was totally misaligned during its surveying and construction and has a conspicuous dog leg in it.

This caused problems as, when a boatman approached the tunnel he could not see if any traffic already occupied the thoroughfare. There are many tales of fisticuffs erupting over which boat should retreat after having met inside. The canal tunnel remained in use up until 1946, after various strengthening works had been carried out with galvanised sheeting in certain places.

A visit was made to the tunnel by Shropshire Mining Club in February 1966. The journey through was said to be very uneventful, except when passing the seven air shafts that are spaced along its length. The airshafts are approximately 40ft in depth from the surface to canal level and a constant flow of water down the shaft walls ensures a wet trip. The water flowing down the shafts has also brought much silt, which constricts the depth of water in the canal to as little as 1ft instead of the usual 4ft. Most of the tunnel is in good condition apart from a collapse of masonry just inside the western portal. It is also stated that the tunnel is host to massive calcite flows, which are similar to formations found in natural caves and are produced by the calcium leaking out of the lime mortar used in the brickwork. The tunnel is still open.

Another canal of interest is that of Donnington Wood Colliery, which is shown on a plan from 1788. The Revd J.G. Cartlidge made reference to it in 1915.

The most striking feature of the mining operations was the existence of an underground canal with boats operating upon it, on what was described as the 'navigable level.' The special shaft with a bell-shaped bottom, down which the boats were lowered on end and then guided out in the bell to rest on the water in a normal position, exist in close proximity to the parish church of Wrockwardine Wood. Some of the older pits discharged their mine water into specially constructed tunnels, which carried the waste water to the underground canal and at the terminus of the canal were fire engines which brought up the surplus supply of water. Sadly, the canal is now inaccessible.

It is very surprising how many mining ventures were linked to underground canal systems. Another underground connection was made at Lilleshall, where one of the limestone quarries was linked to the Donnington Wood surface canal. The two canals join together near a place called Hugh's Bridge; the limestone quarry surface canal however was found to be 18m lower than the main canal. To overcome this, the branch canal travelled beneath the main canal in a tunnel and the two were then joined together by a vertical shaft. A crane then lifted pallets of stone from the boats below and transferred them to the barges above. Many of these features were taken from the Duke of Bridgewater, who was first to exploit underground minerals by water canals at his extensive coal mines at Worsley, Manchester. There are many other features similar to this in the Ironbridge Gorge.

A similar tunnel to Erdington is found at Stableford, where its main use was for transporting large volumes of water from one point to another. I shall now go into this in some detail.

Stableford Tunnel

The tunnel first came to my attention upon reading an old Shropshire Caving and Mining journal from the 1970s. Further information was added in late 1996 after corresponding with Mr Dave Adams, the founder member of the club. He told me how he and a caving mate, Mike Moore, had studied the open ends sometime in January 1989, but knew of no attempt to traverse it by the Shropshire-based club or by other persons. He said that theoretically it should not prove difficult and doubted if any person had traversed the full length of the tunnel for many long years. A further reference was gleaned from written literature, which quoted a similar phrase. 'It is doubtful if anyone now living had traversed the whole length of the 'foot rid' (tunnel) though several claim to have penetrated some distance from either end.' The tunnel was excavated in 1731 and funded by the Davenport family, who lived in nearby Worfield in a splendid period mansion.

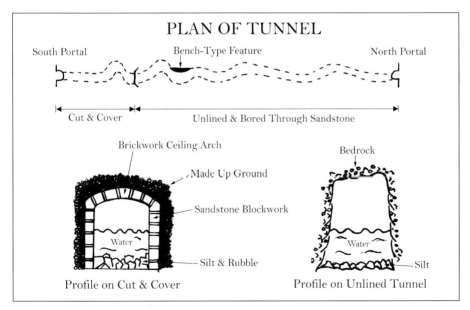

Plan of Stableford Tunnel.

View inside tunnel.

The purpose of the tunnel was to send large volumes of water through the red rock hill to irrigate some 100 acres of ground on Cranmere Farm. It is said that construction costs to complete the tunnel and associated weir and stop gates were very high. The Davenport's spared no expense and held a public sheep roast (barbecue) to celebrate its opening.

The tunnel was used sporadically until the turn of the nineteenth century when it fell into disuse. It was not until after the First World War that it became reused when it was repaired by the tenant farmer of Cranmere. The system was short-lived and by the start of the Second World War in 1939, dereliction had set in and has remained so to this date.

On 2 January 1999 I gained permission from a local farmer to look at the tunnel and assembled, with camera and torch, at its southern portal near to Sheeps Walk Coppice. The portal was little more than 2ft high (60cm) due to the build up of silt at the entrance.

A peep into the tunnel with the torch showed a very low-level tunnel half filled with water. A sheep's skull stared up through the water at me. Undeterred by this, the compulsion to attempt the tunnel overcame me and I squeezed through into the water, which immediately overflowed my wellington boots. The tunnel was so low it left me in a stooping posture. I then proceeded up the first section of tunnel, which was constructed by the cut and cover method, large sandstone blocks made up the vertical walls and red brick arching formed the roofline. The tunnel proceeded for some 50m in a straight line before bending first left then right before straightening out into an unlined tunnel carved into the solid sandstone. I was now able to stand upright in a tunnel profile of 6ft high by 3ft wide at the waterline and 2ft wide at the ceiling, which was mostly flat. The water level never dropped below 2ft 6in in depth (Dave Adams mentioned that only 6in of water was found in the tunnel entrances in 1989).

The tunnel continued in a series of S bends with the roofline dropping sharply in places, leaving me again stooping low towards the water. Harassment by subterranean arachnids and large webs spun across the tunnel width slowed progress marginally. At intervals, carved into the sidewalls were various notches of an A-shaped nature, most of these were close to the waterline, and I conjectured that these were used for holding candles during tunnelling operations.

Steeraway Caves and Tunnels.

Steeraway Caves and Tunnels.

The tunnel also held superb resonant qualities, the splashing noises and occasional bang of the head on the roof echoing nicely down the tunnel. On approaching a last bend, the north-eastern portal was dimly seen. The tunnel appeared to look in a straight line from here on but as I traversed the last section it was noted to meander from side to side slightly. The ceiling height gradually rose until I was once again able to walk upright. A few photographs were taken at the portal and I turned around to traverse the tunnel in the opposite direction. Only one minor problem was encountered on the way back and that was that the water was now a murky brown, leaving me to search for footings as various boulders and branches were submerged on the passage floor. The airflow was noticed to be very good throughout the tunnel, the pick marks made by forging the passageway were still in evidence and there were little signs of erosion or spalling of the roof, considering the age of its construction.

The return trip took a brief five and a half minutes. The tunnel must be 200-250m long, the deepest point under the hill approximately 20m, with the cut and cover section a modest 2-5m below surface level.

Near to the Wrekin Hill is a place known as Steeraway Woods, where there are cave-type features in the carboniferous limestone. Unfortunately, these are of mining origin and are the remains of run in adit tunnels. My son and I explored an unlined portal and tunnel, but this only travelled a short distance. The local lads showed other entrances to us, but these turned out to be the connecting tunnels to a battery type lime kiln associated with the extensive limestone quarrying that took place here. The tunnels are 24ft long and brick-arched and are blocked by rubble at the internal ends. The limestone, I am told, was extracted from 1700 until approximately 1850.

There are also cave-type features found in the sandstone outcrops around North Shropshire, which are of mineral extraction origin. The first is at Pim Hill, where an entrance drops at a steep angle down into the sandstone embankment and quickly becomes a low crawl through infill material, before breaking out into a handsome rock cut level still showing the pick marks made from excavation. The level travels forward before splitting into two passages that both peter out. It is said that these features date back to 1643. During the reign of King Charles I, the entrance was buried until a member of Shropshire Caving and Mining Club dug it out. Another cave-type entrance can be found in an escarpment at Erdiston SJ 366246, this travelled to the now filled shafts and workings that were reputedly made for copper around 1800.

A further place of interest is Queens Head. Michael Raven, in his gazetteer of Shropshire, mentions that three and a half miles south-east of Oswestry, in the fields behind the green corrugated iron sheds that stand beside the canal, there were sand pits. The sand was transported in wagons drawn by donkeys, which hauled it through a low tunnel that emerged in the middle of the shed.

Shrewsbury itself is situated on the sandstone belt and is also no stranger to delving beneath the ground. Unfortunately, being a built up area, the old workings had a detrimental effect on the parish church of St Chad. The present church was built in 1788 after its predecessor sank into the earth with so little noise that no persons nor even the town's watchmen were alarmed or aware of what was happening. Subsidence from old workings was believed responsible.

There are other chambers of antiquity that are still in existence beneath the town. An article was written in the magazine *Current Archaeology* in September 1998 by N. Baker concerning these old, but fascinating, cellars.

Icehouse Structures at Woodcote, near Newport

An area near to woodland away from Woodcote Hall contains a former sandstone quarry. After quarrying had been completed, an icehouse was excavated. The icehouse is quite exceptional as it is completely cut out of the sandstone bedding; no brickwork has been built into the structure as is the case at many other sites (other than the flooring of the entrance chamber). It was also noted that the ice chamber was of an elliptical nature and not circular as in most icehouse construction. The sides are also nearly vertical. A fault line passes through the doorway and entrance chamber and this has weakened the cambered ceiling slightly, while a block of stone had fallen out of the roof at some time past. The stone doorways have been rebated to receive wooden doors and frames, but these are no longer in situ.

There are other features carved out of the quarry face, the first to the right of the icehouse is that of a bench or rest area in a semicircular shape, complete with footrest. This was roofed over with the sandstone until a tree had grown through the roof splitting away the cliff face. A date of 1881 was found carved into the bench. To the left of the icehouse a curious tunnel has been forged through the rock, now bricked up at one end, and much carving of names and dates were found: 1854, 1858 and 1912 as well as more modern ones of 1963 etc. The tunnel is 6ft high, 3ft wide and 10ft long.

At one time the tunnel appears to have had some form of building constructed by its one entrance, as holes for beams could be seen in the cliff face, as well as a vertical notch for a post. The bricked-up doorway is louvered to vent the tunnel and was obviously used as a form of store at some time.

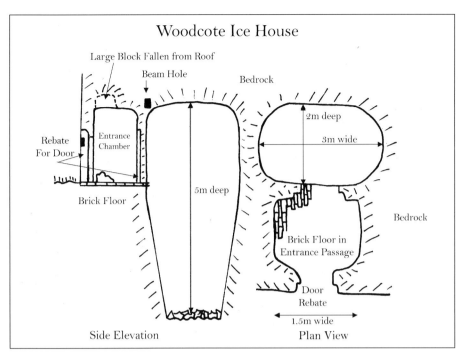

Woodcote Ice House

Large Block Fallen from Roof

Beam Hole

Bedrock

Rebate For Door

Entrance Chamber

2m deep

3m wide

5m deep

Brick Floor

Bedrock

Brick Floor in Entrance Passage

Door Rebate

1.5m wide

Side Elevation

Plan View

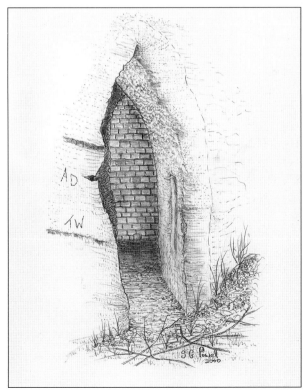

Above: *Plan of Woodcote Icehouse.*

Right: *Drawing of arched tunnel, Woodcote.*

Icehouse entrance, Woodcote.

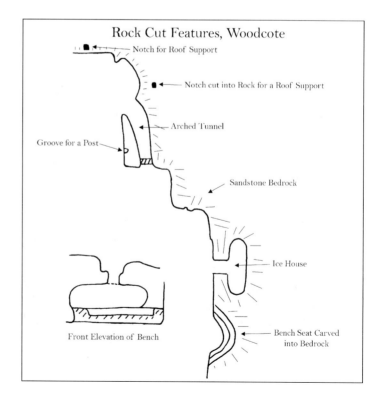

Rock Cut Features, Woodcote

Notch for Roof Support

Notch cut into Rock for a Roof Support

Arched Tunnel

Groove for a Post

Sandstone Bedrock

Ice House

Front Elevation of Bench

Bench Seat Carved into Bedrock

Plan of rock cut features, Woodcote.

44

Davenport's Icehouse, Worfield (Hallon Mere)

This was visited by my son Lee and myself on 30 December 1996 and was found in an embankment close to the road leading down to Hallon Mere. The icehouse was found to be bricked up where the entry passage travels into the hill. There was a brick-fronted porch entrance, of which the roof had collapsed or been demolished, leaving a semicircular retaining red brick wall enclosing the entrance portal. The ground level is substantially higher now, creating a drop into the small entrance passage.

Icehouse entrance.

Ice House at Hallon Mere

Conjectured Chamber

Blocked-up Doorway

Ceiling Line to
Brick Entrance Vault

Soil/Backfill

Soil

Plan View

Front Retaining Wall

Brick-Vaulted Ceiling

Soil & made up ground

Front Retaining
Wall

Conjectured
Chamber

Soil Blocked-up
Doorway

Side Elevation

Plan of Hallon Mere Icehouse.

The icehouse belonged to Davenport Hall and was built between 1726–27 by Henry Davenport, the house gardens and icehouse being funded by a fortune amassed by him in India, during his younger years. The ice used to fill the interior was taken from the large Hallon Mere, found approximately 1km below its entrance. The icehouse is north facing to minimise the likelihood of warm air blowing in the direction of its doorway during the spring and summer months. It is said that a small passage leads from the now bricked-up portal to an egg-shaped icehouse constructed from brickwork, which has the large depth of 20ft.

The icehouse was bricked up some years ago after a disastrous situation occurred within. A local tenant farmer had the misfortune to find that a boar and nine sows belonging to him had wandered into the underground passage and had fallen into the interior cone. The noise given by the poor animals alerted nearby workmen who, with the farmer, went to investigate the situation. A torch beam showed the animals huddled together and none the worse for their fall into the depths. At first they tried to haul the pigs up by rope, but the pigs continually fell through the noose. The smell was getting unbearable for the men working in the icehouse, and the animals began to get distressed, so everybody retired to rest for the night. The farmer was unable to settle and thought up a simple method of getting the pigs to the surface by building up the floor level with bales of straw until the ground level was reached where, at this momentous moment, the pigs happily trotted out.

Mossey Green Icehouse

A Mr Woolley of the nearby Clive Cottage showed me the icehouse, or subterranean structure, at Mossey Green on 14 July 1996. He kindly took the time to drop what he was doing, put on a hat, coat and boots, so that he could show my son and I the precise spot of the structure, as I had found difficulty in trying to locate it myself.

His help was very much appreciated. Mr Woolley also told us that he had heard local rumours of a tunnel linking the underground structure to the nearby mansion of William Reynolds. The mansion was originally built in 1729 and, although Reynolds was still operating as an industrial entrepreneur around Telford and Ketley in the 1780s, building secret tunnels was not one of his pastimes.

Mossey Green Vault.

Underground Structure at Mossey Green, Ketley, Near Telford

Plan View

Ground dug away exposing covering spoil to dome

A

Part collapsed tunnel

Blockage of tunnel

A Direction of further stone structures

Line of collapsed tunnels

Former shaft opening in top of dome, bricked up at a later date

3m

1m width

Todays main entrance

Large block cast concrete

Corbelled shaft top

1m

Main Dome

Section on AA

5m

3m 1.5m

Tunnel 500mm

900mm high

1.5m thick

Access tunnel

Rubble filled to here

1.5m

Conjectured outline of cone

3.5m

Outline of main entrance

Plan of Mossey Green Vault.

The Mossey Green structure has a domed roof made of cut sandstone blocks. There was originally an access hole in the centre of the dome, possibly as a vent or for lifting materials in and out of the centre of the dome. This had been blocked up at a later date in time. There are two long connecting tunnels that appear to the sides of the dome and a third opening set in the middle of the dome. A previous excavation had dug away the covering spoil to the rear side of the chamber, where some of the brickwork/blockwork had been damaged. The interior is very similar to the icehouse cones or egg-shaped chambers seen at other places around Shropshire, except that the chamber is only 10ft deep.

This diminishes towards its base from the top, like other icehouses, from a diameter of 9ft to 4ft, and at the bottom of the dome is a cast-iron perforated plate, obviously for drainage. However, other theories for its use were stated in 1976; a potters kiln, a smelter, a steam engine boiler and last but not least, a Dundonalt prototype kiln for extracting tar from coal. My own synopsis gave rise to possibilities of it being a malting kiln, similar structures being in evidence in the Nottingham caves.

The structure was first investigated in 1973 and was located by surveyors on the line of a new road; local boys at the time called it a cave. Luckily, during March 1975, the Telford Development Corporation announced that in view of its interest and antiquity, the line of the new road would be moved 45ft to safeguard the structure for future generations.

Beckbury

Lee and I visited this building on 3 January 1999. On the top of a hill to the rear of Beckbury is a manmade amphitheatre. First to be seen is a raised platform, reputedly used for fighting cocks in the unruly eighteenth century. Nearby is found a large square stone slab with an attached iron ring – removal of this very heavy slab uncovers an underground chamber of small dimensions, half filled with water and obviously some form of well. Another underground chamber is found near to the cock pit, which, according to Vivian Bird in *Exploring the West Midlands*, was called an icehouse.

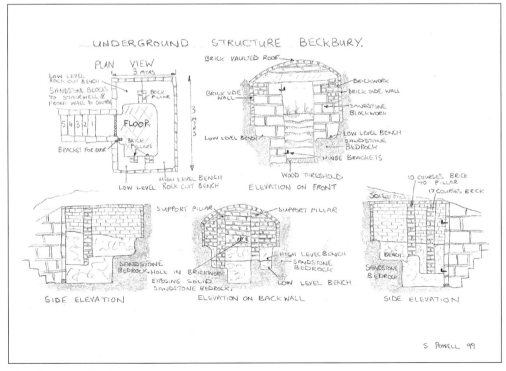

Plan of Beckbury Chamber.

Beckbury Chamber.

On closer inspection, the structure is more of a cold store than an icehouse and would have only been a temporary measure for storing perhaps poultry and game or beer for the fights. Only the bottom half of the chamber is carved in sandstone, the top half is built of red brick, including a vaulted roof, which has been added later and covered over with soil.

There are six stone steps down into the chamber. A raised bench of some 50cm has been carved out of the sandstone for storage on three sides, this bench steps down towards the door on the two sides. For unknown reasons, hinge brackets for a door are found in the brickwork, indicating an inward opening gate. Three brick pillars are in position as supports for the vaulted roof.

Tong, Patshull, Lilleshall and Haughton

At Patshull Hall and pool, an icehouse structure existed until the mid-1960s, when it was unfortunately demolished. It is said that the building was of a cylindrical nature, part sunk into the ground. A thatched roof had been added to give the building character. It was last used *c.*1910 and was positioned in the southern corner of the walled garden.

Another unusual icehouse exists at Lilleshall. The icehouse is brick built and rectangular, instead of the traditional egg shape, and it has a brick-vaulted roof. It is positioned in the southern bank of a pool found in the golf course, and is a good half-mile away from Lilleshall Hall, which it once served.

Haughton Hall, built in 1718, had the luxury of two icehouses, one subterranean, used to store ice on a long-term basis and a small recess store nearer to the house, the ice being taken from one to the other and stored as a temporary measure for use in the Hall.

They were last used for their original purposes in the mid-1900s. The icehouse is situated on the lake near to Wesley Brook outlet and is buried in the bank. Its interior was of the traditional domed egg shape and it is unfortunately bricked up.

Burlington House Farm has two vaulted structures, approximately 12ft by 6ft, whose domed roofs are just below ground level. Ice was taken in winter from the pool in front of the farmhouse and stored in these chambers, it was then used for cooling milk in later months.

The Ordnance Survey map of 1954 shows an icehouse existing in the grounds of Apley Park near Bridgnorth, situated on the outskirts of wooded ground 200m north of the Stately Apley Hall.

Within East Shropshire is the village of Tong. Nearby are the scant remains of Tong Castle. This is host to an icehouse built in the eighteenth century which was of very large size: 3.5m in diameter by 5.8m deep. It is of the traditional egg shape. Ice was taken from the large south pool that formerly existed nearby, although the north pool in front of the castle ruins is still in existence today. The icehouse has a small side entrance which, along with the top of the icehouse

Entrance to tunnel at Tong Castle.

dome, lay beneath ground. A tunnel linked the icehouse with the castle cellars, where the ice could be prepared for cooling wine and foods. The entrance to the conjectured house was accessed by myself in 1996 and a small arched tunnel of cut sandstone blocks led to a bricked-up square portal that would have had access to the icehouse dome.

It is impossible to say if the dome is intact inside as it is positioned on the edge of the M54 motorway which had cut through the grounds of the castle and, to keep the embankment stabilised, the chamber may have been filled in or destroyed. There is an article that mentions an icehouse from Tong being removed and rebuilt at Avoncroft Museum. In the book *The Wandering Worfe* by D.H. Robinson, he states that there are two icehouses, one near the base of the lake and a further one 100-200m downstream. This leaves us with the slight problem as to which icehouse was reconstructed.

Badger Dingle

On the outskirts of Badger, in the grounds of the old Badger Hall (now demolished), a series of underground cavities can be found. The first (No.1) is of a small alcove, cut back into the rock face approximately 1.2m in depth by 2m high by 1.2m wide and presumed to have been a shelter out of the suns rays or resting point for persons admiring the garden walks.

No.2 is an icehouse built for the hall in 1837. The icehouse is of the very basic design of entrance tunnel and inner and outer doors leading to an egg-shaped chamber. The bottom of the dome has been constructed from brick. There are two strange slots excavated in the top of the icehouse pit, which form an arch type feature.

Badgers Dingle.

The next items of interest are caves (3) and (4). These appear to have been formed originally by the backwash of the stream and have now been left behind on the escarpment 15ft above the streambed. Cave No.3 however, has been adapted for other purposes, due to the guiding of water into hewn troughs, which must have been used for purposes of the garden walk. Further downstream we again come to a natural rock shelter formed by water erosion; this again was utilised into the walk as a resting point with a bench for seating. Just below this point a small aperture was found and access was gained into a cave 30ft long by 5ft high and 5ft wide, the entrance much clogged up with rotting vegetation. The cave was also habitually visited by a fox, one corner being left as a dump for bird carcasses after its feasts.

At stream level a brick-fronted portal was found traversing back 3m to a bedrock back wall. An offset brick shaft was formed to the left-hand side of the tunnel 10ft high by 3ft by 2ft 6in, the purpose of this structure is unknown. Another strange tunnel structure was found at stream level, this time travelling in the same direction as the stream. Again purposes not known other than for diverting the stream. The last features at Badger Dingle are two small caves of natural origin and a large tunnel excavated through the sandstone ridge to connect the bottom level of the walk to the top path. The tunnel is approximately 40ft long by 7ft wide by 8ft high.

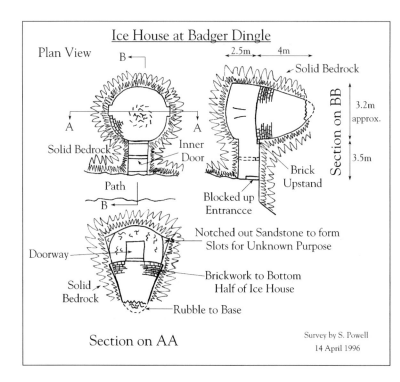

Ice House at Badger Dingle

Plan View

2.5m 4m

Solid Bedrock

Section on BB

3.2m approx.

3.5m

Solid Bedrock

Inner Door

Brick Upstand

Path

Blocked up Entrancce

Notched out Sandstone to form Slots for Unknown Purpose

Doorway

Brickwork to Bottom Half of Ice House

Solid Bedrock

Rubble to Base

Section on AA

Survey by S. Powell
14 April 1996

Left: *Plan of ice house.*

Opposite: *Plans of chambers at Badgers Dingle.*

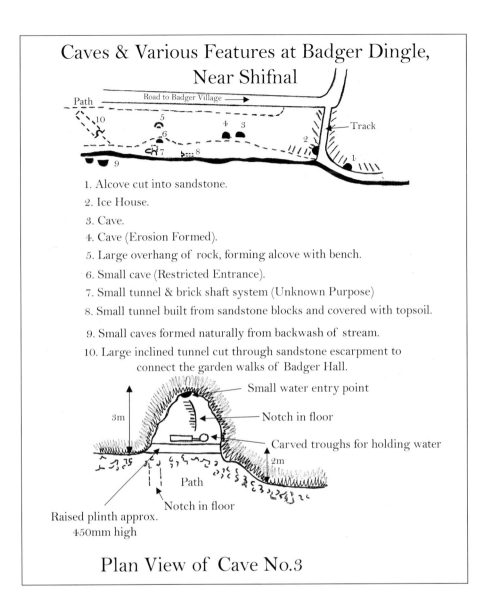

Caves & Various Features at Badger Dingle, Near Shifnal

Path — Road to Badger Village →

Track

1. Alcove cut into sandstone.
2. Ice House.
3. Cave.
4. Cave (Erosion Formed).
5. Large overhang of rock, forming alcove with bench.
6. Small cave (Restricted Entrance).
7. Small tunnel & brick shaft system (Unknown Purpose)
8. Small tunnel built from sandstone blocks and covered with topsoil.
9. Small caves formed naturally from backwash of stream.
10. Large inclined tunnel cut through sandstone escarpment to connect the garden walks of Badger Hall.

Small water entry point

3m

Notch in floor

Carved troughs for holding water

2m

Path

Notch in floor

Raised plinth approx. 450mm high

Plan View of Cave No.3

One of the local people mentioned to me that he had entered another ice house by crawling into a small silted chamber, connected by a passage which led to the usual bell-shaped ice pit. Another article mentions an icehouse at Badger with a steel tank inside. Neither of them was found on this occasion.

The main icehouse is approximately 180 years old and was in use in connection with Badger Hall up until the 1930s. A most interesting account, and the only one I have come across of the procedures used to store ice, is given by D.H. Robinson in *The Wandering Worfe*.

Stapeley Hill

Stapeley Hill is well known for its cave because it has always been included as part of the surface features on local maps. They have always termed it the 'Giants Cave'. There does not appear to be any source material in 'folklore' or in local rumour to suggest why such a name is given. A summer walk round the site led me to believe that it had magically disappeared or collapsed and I wondered how many other ramblers or historical researchers had come to the site out of curiosity to end up in bewilderment, wondering what fate had dealt to the cave in question.

My inquisitive nature led me to find an old trip report of Shropshire Caving and Mining Clubs dated from 1961. It mentioned how the farmer, who inhabited the cottage below the cave site, told them that the cave existed until the early part of the twentieth century and was blocked up to stop cattle wandering in. He told them that the entrance was believed to be to the left of some large boulders behind a holly bush. The cavers commenced a dig at this point but quickly came to solid rock. A further dig was made nearby that was also unsuccessful, although the ground in the vicinity appeared to have been disturbed at some time past. At this point the search was abandoned until they could gain fresh information.

The former owner of the cottage at Stapeley Hill was found alive and well, living in nearby Wellington town. He was now eighty-five years old and said he could still remember the exact details of the site. He stated how two calves had strayed into the cave and had died of some unknown disease; the cave was then blocked up. 'I was seven, maybe eight years old at the time', he said. Giving a date of around 1884-5 for the cave's closure, he said that the cave mouth led in underneath the tall boulders. The cave was artificial and not more than 20ft long. When it was blocked several very large boulders were rolled into it and several feet of rubble was filled in on top, making the level area that exists today. It was considered that it may have been an ancient mine, as the rock is too hard to have formed a cavity naturally. Other rumours exist that it had been inhabited by prehistoric man, but neither idea can be proven and it was felt that nothing could be gained by re-opening the cave, due to the foreseeable difficulties of the large boulders now lying inside the cave. There was, however, much activity in the area by Stone Age people; nearby is Mitchell's Fold Stone Circle and one of the valleys was used as a Stone Age factory for making axe heads.

Caer Caradoc

Having broken in the millennium year with a few beers and no sleep, I decided the best way to start the new year was by getting a bit of fresh air, by walking the Church Stretton Hills, which are also host to a place known as 'Caractacus Hole'. It is positioned on the steepest face of the hill and is at SO 4795. Georgina Jackson, in 1883, states that, 'This is where the people declare the king hid from his enemies after his defeat. Caractacus being hunted by Roman Legionary'.

The cave is positioned a few metres below the ramparts of a large hill fort and there is nothing to suggest whether it is a natural or a manmade cavity. It may well have been a trial adit for mineral prospecting, a shepherd's shelter, or even a lookout post, we shall never know. The cave was, however, quite old when G. Jackson wrote of it.

The cave is now logged by the Department of the Environment and is listed in the 1978 edition of *Ancient Monuments of England*.

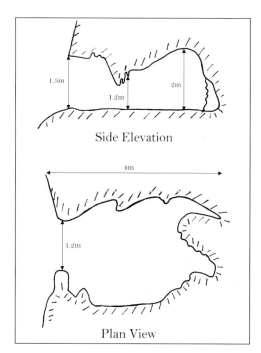

Plan of Caradoc Cave.

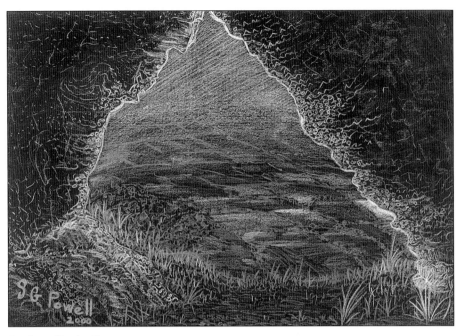

Interior to Caradoc Cave.

Exterior of Caradoc Cave.

Caynton Temple

Near to the village of Beckbury, we find Caynton Hall at SJ 775029. Caynton Hall was built in 1780 and on its garden fringes, some 150m away, we find an area of very disturbed ground that may have formerly been host to small quarry workings. A sign stating 'Danger Subsidence' has to be passed on the way to what we can only describe as a rock cut temple or grotto. There is much information appertaining to the hall and its occupants but, unfortunately, the temple is a complete mystery.

There are rumours locally that a secret passage links the temple to Caynton Hall, but this is sheer speculation. It is possible that the area was once a garden grotto feature attached to the hall, as it was fashionable at that time to have small underground features. A narrow slit-like entrance drops down through the side of an embankment for approximately 2m to enter a short passageway that leads us into the main chamber. The author Vivien Bird in her book *Exploring the West Midlands* makes the mention that, 'The entrance a few years ago was only a mere fox hole and it is now much enlarged.' She also mentions the subterranean temple being built by General Legge, who lived in Caynton Hall.

This means that the original entrance is through one of the backfilled levels. This would place the visitor directly in front of the central pillar construction, this large pediment being the focal point to the chamber. Its base is a single cylinder approximately 1m high by 1.5m in diameter, which forms a table-like plateau from which four small pillars spring. These are capped by a flat slab from which the roof arches out in all directions. This central configuration has been used for small fires on occasions, to light up the chamber by youthful explorers.

The whole complex is built to ecclesiastical standards with bench pews, alcoves and fancifully carved niches. One such alcove is sprung by four arches, carved with studs. The focal point to its centre is a carefully placed niche for a light or religious cross. Its whole design reminds me of the Norman era. The walls, at one time, were decorated with coloured pebbles or shells, as the reliefs of flowers and circles can still be made out, the decorations having long since been picked out of their bedding.

Left and opposite: *Caynton Temple.*

One can quickly imagine how striking the temple would have been in its wholesome state. There is a far anti-chamber, a step down leads into the circular room, and its periphery is carved into dummy support pillars. Around its central area on the floor is a carved bowl or urn and, being made from sandstone, was obviously not for holding water. A look at the plan layout of the temple gives an impression of the map of Great Britain. This, however, may only be my own imagination being inspired by this wonder of Shropshire.

Unfortunately time is beginning to tell on the structure, as youths carve names on the walls and cut through support pillars. The original pick marks are still in profusion through the graffiti. It is a shame that heritage groups can only make preservation orders on structures above ground. English Heritage is only able to list an entrance of subterranea if it is of architectural interest, yet this place should surely be gated and saved for future generations to enjoy.

It is unclear if an extension existed to the temple, as backfilled levels lead from the area of collapsed/disturbed ground.

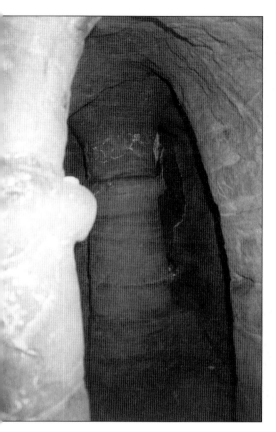
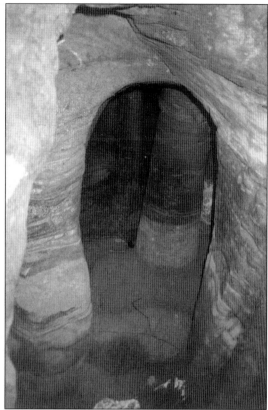

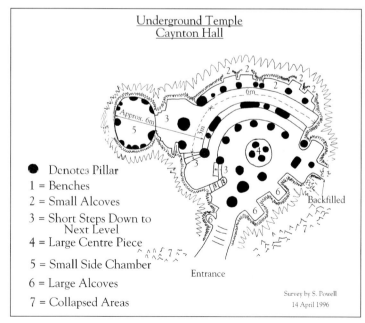

Underground Temple
Caynton Hall

● Denotes Pillar
1 = Benches
2 = Small Alcoves
3 = Short Steps Down to
      Next Level
4 = Large Centre Piece
5 = Small Side Chamber
6 = Large Alcoves
7 = Collapsed Areas

Survey by S. Powell
14 April 1996

Plan of Caynton Temple.

Decker Hall Caves

An early Shropshire Caving Club journal quotes there being at Shifnal ornamental caves in sandstone, about 6ft high, the inside being damp and covered in green moss. These are locates at SJ 750095.

Another reference states that there are caves in the River Worfe area. Caves of various sizes, some suitable for dwellings, exist at Shifnal golf courses. These particular caves are now used for breeding wild fowl and are unfortunately off limits.

The Shropshire Caving and Mining Club journal of 1979 ran an article on tunnels at Decker Hill. Within the grounds of the hall are two underground manmade arched tunnels, used as secluded roadways through the estate by the lords of the manor. They are both estimated to be about 18m long, and were built on a curve to discourage livestock from wandering through. Their construction is probably c.1810, with the house dating from 1784.

Near the tunnels is a possible icehouse, although it is only accessible by means of ropes. The tunnels are situated near to Decker Hill Farm. Immediately opposite the farm, in a wall at the junction of two lanes, is one of the entrances to tunnel no.1. This runs under the farm lane and is apparently blocked up at the other end. Continue straight on past the farm and along a high brick wall to the woodland beside the lane.

The second tunnel, which is open at both ends, runs from the wood under the farm track. Beneath the hall itself are a series of brick built drainage tunnels, which are large enough to crawl through. Cast iron drainage pipes appear to have been installed at a later period.

A visit was made to locate the ornamental caves and tunnels but, unfortunately, the landowner was most hostile and uncompromising, so we will have to make do with the information given already.

Tong Forge Pool

It is quite possible that the dwellings excavated here are of substantial antiquity, as the area was licensed as a forge by the Earl of Shrewsbury in 1564 and it is reputed that it was still working in 1806. It is quite feasible to assume the rock house was used in connection with the forge. The cave house was habitable until 1832 when disaster struck. A large flood occurred, washing away most of the brick frontage. The rock house was replaced by a substantial building on the clifftop, away from any contact with the water below. The intervening years have since seen the rock house caves used as wood stores in conjunction with the house above.

View of Tong Forge Pool.

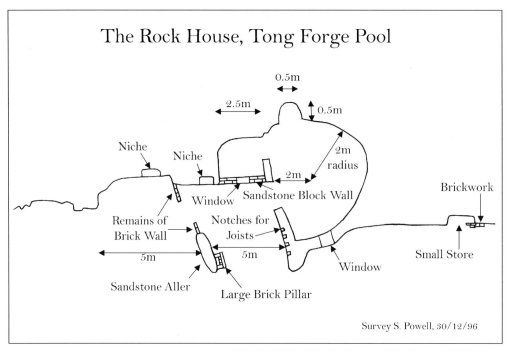

The Rock House, Tong Forge Pool

0.5m

2.5m

0.5m

Niche

Niche

2m radius

2m

Brickwork

Window   Sandstone Block Wall

Remains of
Brick Wall

Notches for
Joists

Small Store

5m   5m

Window

Sandstone Aller

Large Brick Pillar

Survey S. Powell, 30/12/96

Plan of Tong Forge Caves.

The kind owners allowed me to inspect the rock house caves on 30 December 1996. The dwelling still showing brickwork and footings from its 1832 era. The main underground cave chamber was still in excellent condition. This is of a semicircular nature with plenty of headroom, approximately 3m. A small window is set into the sandstone outer wall with part of the cave wall enhanced with red brickwork to the side of the main chamber. A small store cave has been excavated out of the bedrock 1.5m high by 0.5m in depth and width. To the left of this store is a doorway into a chamber of small dimensions, 2.5m long, 1.5m wide by 2m high. The front to this chamber is made by the addition of a wall made from sandstone blocks. There is a window aperture in place. I have often wondered if these have been added since the great flood to utilise greater storage space. Niches carved into the sandstone cliff are still in situ, giving evidence of two further large rooms to the complex. Due to the shape of the main cave, it is possible that it may have originally been broadened in size as the rooms and walls were added to form it into a dwelling place.

To the right-hand-side of the rock house is a small store, the aperture only now survives along with a handful of decaying bricks. The cliffs proceeding the rock house are well undercut by water erosion, giving further evidence to support the rock house's natural start.

Higford Rock Dwellings

There are at present two cave-type structures, one of which is used for a coal store by the nearby cottages.

The larger cave is empty and has not been used for any purpose for a number of years, due to collapses of the cliff face, which has already shortened the cave in length. It was difficult to imagine the caves having been used for any kind of living accommodation.

The main cave is 3m high by 3m long by 2.1m wide with an alcove on the left-hand side. The coal store is 1.2m high.

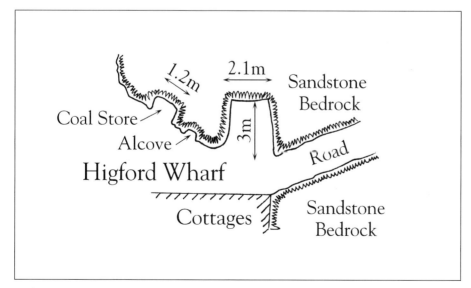

Higford Wharf Caves and plan.

Worfield Rock Houses

The rock houses are situated below the village of Worfield and are excavated in the cliffs next to the River Worfe. There are enough remains of the open chambers to surmise living accommodation for perhaps two families, but this accommodation appears to be very small. It was noted that the mid-section had been altered with brickwork to help support the cave structure at some time and brickwork was evident by the largest cave, suggesting it may have had brick wall fronts. It is presumed that this is the site of St Peter's Well and cave, mentioned in the book *Worfield on the Worfe* by S.B. James, written in 1878.

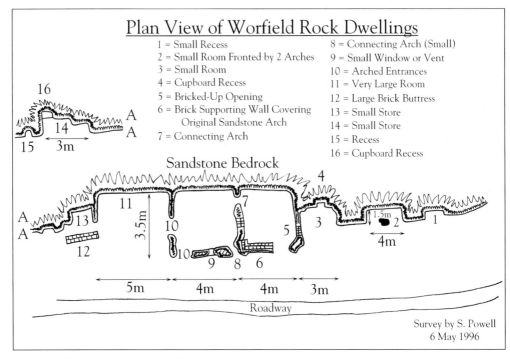

Plan of rock houses, Worfield.

Worfield Rock Houses.

Near the well was a cave which, with a built-up brick frontage and a door and window inserted, served as the home for many years of an old lady named Sarah. She was very much a local character and a keen churchgoer, until one day she suspected the vicar of directing his text 'Wash you and make you clean' specifically at her. From that day onwards she never resumed going to church, the conditions at the rock house being such as to make it difficult to keep it free from soot, soil and sandstone.

Worfield Rock Houses.

A similar cave system is found at Evilith Mill on the Wesley Brook, some five miles distance from Worfield. The mill house dates back to the 1500s and, cut into the cliffs at the back of the mill, are caves. These were used for a variety of purposes; one served as a dairy, one as a bakery and one for washing utensils, the water being supplied to the cave by pipes from the headwaters upstream of the caves.

At SJ 469243 a brook crosses the A528 just north of the village of Middle. This may possibly be the site of a cave house mentioned by G. Jackson in 1883: 'Bristle Bridge, Myddle – there is a certain cave in the rock near this bridge, which was formerly a hole in the rock and called Goblins Hole, and afterwards was made into a habitation with a stone chimney built up into it by one Fardow.'

The earliest signs of any cave dwellings in Shropshire were uncovered during excavations at Burcote Rocks near Worfield at SO 7450 9515. Flint artefacts and animal bones of an unknown date were logged at that time. (This information was supplied by an article held by Sheffield University, entitled *Remains of Early Hominoid Settlements in British Caves.*)

Kynaston's Cave

This is perhaps one of Shropshire's most famous sites and is located at SJ 388192, near to the village of Nescliffe. The cave is situated in the Cliff of Ness, a hill formed of new red sandstone with a south-facing scarp.

The cave was named after Humphrey Kynaston, a constable to the Castle of Middle. Humphrey was the son of Roger Kynaston, his mother the daughter of the Earl of Tankerville and Lord of Powys and he was born in the later half of the 1400s. His fame and notoriety surfaced during the reign of Henry VII when he became short of money and took to highway robbery. His deprivations became so severe that in 1490-1491 he was officially outlawed and was then forced to flee from his dilapidated castle at Middle and find refuge elsewhere. It is said that he wandered onto the cliff and began to cut a long flight of steps into a buttress of the sandstone and, on reaching the vertical cliff face, hollowed out a doorway and two chambers. One of them was to serve as a stable for his horse and the other for his own habitation. His own room was completed with a hearth and chimney flue, a circular window was cut into the living rock close to the doorway for light and especially for observation of the stairway. The rooms are of small sizes, one 11ft by 11ft, the second 7ft by 5ft, the ceiling primitively carved to give a vaulted effect. There are also large socket holes in the bedrock, where a large locking beam would have been placed behind the solid door. The stairway to the cave house was constructed so as to allow the passage of one person at a time. Humphrey could then fend off any attacker with a long pike. At the base of the stone steps is a stone trough cut into the bedrock, this would have served as a feeding vessel for his horse and filled with wild oats or corn.

In the dividing wall of the chambers are carved niches, probably for a lamp or candle and the initials HK 1564. Humphrey died in 1534 and obviously had nothing to do with the signature. As an outlaw he became known as the Robin Hood of Shropshire, robbing the rich and befriending the poor. Humphrey was pardoned on 30 May 1493, the document still existing in 1911 and held by a descendant living at Hardwick Hall and Hordley. Once pardoned, Humphrey moved to Welshpool. The cave was not abandoned and became used in later days by others of a troglodytic nature and it is noted that a family of nine was

Engravings by Baring Gould – Kynaston's Cave.

living here in the late eighteenth century. The cave in modern times is host to a bat population and entry to the interior chambers is restricted and gated. In 1822 the cave was mentioned in the book *Old Stories* by Miss E. Spence. It states 'It is the present abode of an old woman who shows it is a curiosity to the traveller.'

In 1994 a timber flight of stairs was constructed to save any further wear and tear to the existing sandstone steps. At the top of the new staircase a view inside the cave is just discernible.

The cave is mentioned by many authors, not least G. Jackson in 1883 and Revd Baring-Gould in 1911, and the Victoria County History of Salop also makes reference to it.

Since then, the cave and Kynaston have become known in folklore for a ghost rider, occasionally coming forth to frighten some poor local. The cave is also said to have had an iron door which, according to legend, became the door of Shrewsbury Gaol (prison). Kynaston's cave is now classed as an ancient monument and is listed in the 1978 volumes published by the Department of the Environment (*Ancient Monuments in England*).

70

Colemore Green, near Bridgnorth

My son Lee, my father and myself visited this cave on 15 November 1998. Access to a cave carved into the sandstone was made via a public footpath from the main road, which dropped steeply through woodland to emerge in a small meadowland valley (Gorsty Hill), dropping down to the River Severn. The cave is positioned halfway along the valley and, after inspection, was found to be nothing more than a storeroom of sorts.

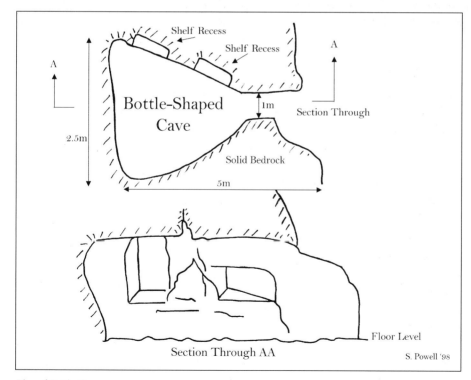

Plan of Bottle Cave.

Opposite: *Plan of Cave House, Colemore Green.*

Left and below: *Bottle Cave.*

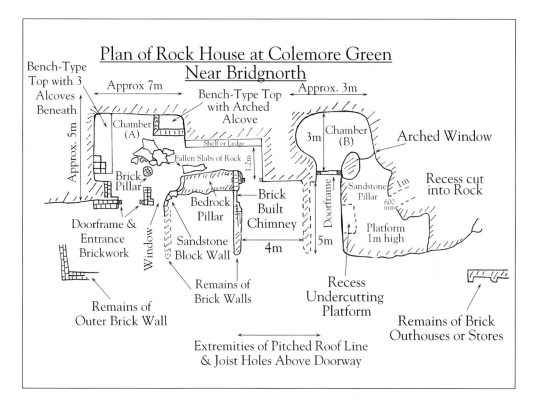

Plan of Rock House at Colemore Green Near Bridgnorth

Bench-Type Top with 3 Alcoves Beneath

Approx 7m

Bench-Type Top with Arched Alcove

Approx. 3m

Approx. 5m

Chamber (A)

Shelf or Ledge

Fallen Slabs of Rock

1.2m

3m

Chamber (B)

Arched Window

Brick Pillar

Bedrock Pillar

Brick Built Chimney

Doorframe

Sandstone Pillar

1m

600 mm

Recess cut into Rock

Doorframe & Entrance Brickwork

Window

Sandstone Block Wall

4m

5m

Platform 1m high

Remains of Outer Brick Wall

Remains of Brick Walls

Recess Undercutting Platform

Remains of Brick Outhouses or Stores

Extremities of Pitched Roof Line & Joist Holes Above Doorway

Its entrance is approximately 2m high by 1m wide. The inside was shaped very much like a bottle, with two recess–type shelves situated on the right-hand wall. Pick marks were particularly evident on the walls. There was a noticeable fault line in the roof of the cave some 200mm wide, out of which sandstone was flaking. The ceiling height was sufficient for standing upright in. The thick mud covering the floor of the cave was produced by sporadic occupation by nearby cattle.

This site was further visited on 15 November 1998, this time by continuing along the public footpath to the river, through knee deep mud, and turning right on the path heading back along the river towards Bridgnorth to a place called Chestnut Coppice.

An interesting cave house or rock house was found, of which two chambers were carved into the sandstone and both of which were accessible at this time. The stone and brick frontages had been removed sometime in the 1960s due to its becoming a haven for tramps. The fireplace to the main living room could still be seen as well as the outlines of the walls. The underground chambers are now beginning to disintegrate, especially chamber (A) where large blocks of stone had fallen from the roof. Large cracks could also be followed from the doorway and window across the ceiling line and I conjecture that this chamber will not remain in the near future. Problems must have been evident many years ago as a steel bar had been added for support at the entrance as well as a circular brick pillar approximately 1m from the doorway, the brick pillar now only having half its original bearing on the ceiling. There was inside a full-length bench constructed from brick and lined on top with quarry tiles with three arched alcoves beneath.

A further bench with an alcove had been constructed in another corner of the room, while the corridor connecting from the chamber to a further doorway had a full-length shelf/ledge carved out of the sandstone.

The second chamber (B) was accessible by a small corridor running between the main house structure and a sandstone ledge of which a recess of 1m long by 0.5m by 0.5m was hewn underneath. The doorway was of very small dimensions, 1m square, this then opened out into a chamber of standing room height, approximately 2m by 3m wide by 3m depth. To the right of the

Colmore Green.

Colmore Green.

doorway, on the ledge outside, an arched window had been carved through the bedrock to lighten up the chamber. A further recess was carved into the rock to the right-hand cliff of the wall for some forgotten purpose.

The cliff face above the chambers showed the line of the original pitched roofs that once existed to the outer brick walls. Joist holes were also present. Further brick structures like stabling or outhouses had existed to the extreme right of the complex.

A further visit was made on 2 January 2000 when it was noticed that vandals had knocked down the supporting brick pillar. An old timer, however, had mentioned that there were caves on an outcrop of sandstone at Rookery Coppice. A search was made on a very unstable cliff-type feature. All that could be found was a very insignificant small fissure cave of 30cm in diameter that travelled along a fault for approximately 2m. Large boulders had fallen away from the escarpment. If caves had existed here, they have now collapsed.

Soudley Rocks, near Bridgnorth

These were visited on 4 October 1997. Access to them was by a beautiful countryside walk along paths running adjacent to the river, starting at Rindleford. The area was heavily overgrown; nevertheless, it was possible to interpret the dwelling areas. There was a large cave-type recess some 4m in length by 1.5m in depth and approximately 2.5m high. Sandstone blockwork was evident on the base of the left-hand corner and in an area 1.5m up the cliff face, to the right-hand side of the recess. The cliff face forms the back wall to the house or cottage. No beam or purlin holes could be seen above the cave in the cliff face for the roof. Some sporadic red brickwork was found on the floor area.

Some 10m to the right, along the escarpment, is a small domed cave with a 2m high roofline with a small doorway of 1m in width. Grooves are cut into the sandstone for the fitting of wooden doorposts, pick marks from its

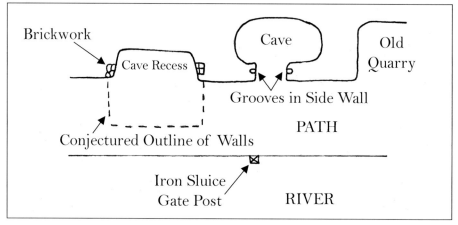

Plan of Soudley Caves.

Cave, Soudley.

construction were evident and the interior had been painted with a white lime wash, which had now faded and mostly flaked away. A wall built into the river edge directly in front of the cave is what might be determined as the remains of some form of sluice gate. By moving approximatley 10m to the right, the remains of a stone quarry can be found perched 3m up the escarpment. It is most probable that the stone for the rock houses came from this site.

The historical notes are as follows:

> *The high sandstone rocks close in the confined but exceedingly picturesque gorge which leads to Rindleford. We at once find deep seclusion and quietude, a scene full of poetic beauty, an old mill on the estate, a sandstone bridge and a cave dwelling in a lofty precipitous rock, inhabited by a lone woman, are a pleasing link to the busy world we have just left.*

This was written by Samuel James over 100 years ago. The sandstone bridge is still in existence some 100m away from the rock house and the old mill I presume to be the one at Rindleford, about a kilometre away. The rock house was inhabited into the early 1900s, later documents referring to it as 'the cave' rather than a cottage.

Winscote Hill
Rock Houses

It is not known when the dwellings were last occupied, other than that they must have been excavated around 1804, the date of much construction of access roads and the nearby Hunters Bridge. The base of Winscote Hill was to become a small wharf for loading and unloading boats. It is quite possible that the rock house was used for overnight stops, during the period the wharf was in use, similar to the rock houses at Gibraltar in Kinver. My son and I were to visit this dwelling on 31 December 1996 and were much dismayed at the way the elements had eroded them. There was one small chamber intact and this has a very small window to the side entrance and also a very low ceiling only 1.5m in height. Access could be made under a low archway or roofing pillar, before entering into what would have been another room. The corridor then continued to a third room, passing by another roofing pillar. The benching carved out of the solid sandstone suggests that this room may have been a cold meat store or some kind of general storeroom. At a lower level, nearer to the river was another small alcove or store. There was a lot of debris and boulders piling up from collapses of the sandstone outcrops and from the frontage of the rock house.

The area is now fenced off and danger signs have been erected due to the cliff's instability.

Winscote Caves.

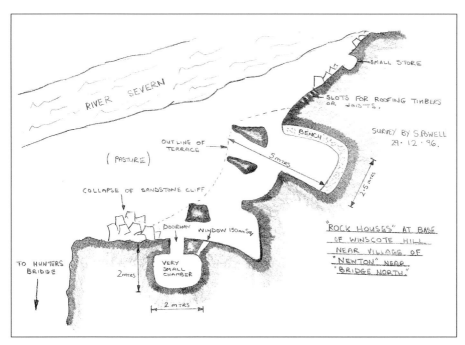

Plan of Winscote Hill Caves.

Cave Cottage, Apley Park

Cave Cottage was found in a large entanglement of ivy and brambles and it is probably totally hidden from view during the summer months. The only reason I managed to find the cave house was due to the kind help given to me by the Apley Estates gamekeeper. His family and ancestors have worked the terraces for the past one hundred years and he knows every nook and cranny.

It was impossible to find the remains of any brickwork or structures that may have encompassed the cave house due to the overgrowth. A small section of the cavern roof has collapsed along with part of a sidewall of natural sandstone; this has left one room part open to the elements. There are many cracks appearing in the cave, showing its present state of instability. However, it was still possible to access and survey the cave as it had originally been.

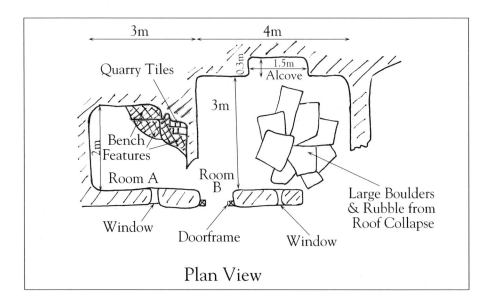

Plan View

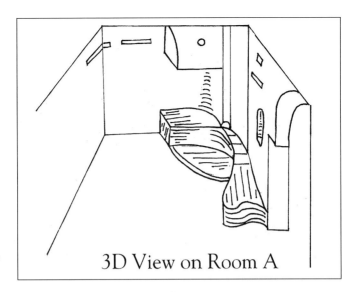

3D View on Room A

Right, below and opposite: Plans of Cave Cottage, Apley.

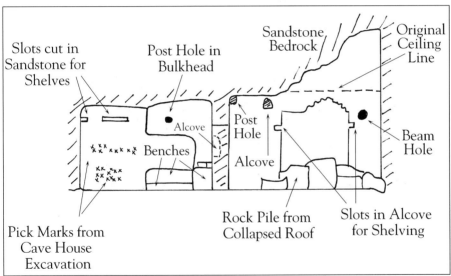

Sandstone Bedrock

Original Ceiling Line

Slots cut in Sandstone for Shelves

Post Hole in Bulkhead

Alcove

Post Hole

Benches

Alcove

Beam Hole

Pick Marks from Cave House Excavation

Rock Pile from Collapsed Roof

Slots in Alcove for Shelving

The pick marks were still clearly visible from its excavation and the cave walls still showed the coatings of lime wash used for its decoration. There were slots cut into the sandstone for brackets and shelving. An interesting feature by the doorway is a notch, possibly for a candle or lantern. There are also bench–type features with a quarry tile covering. The gamekeeper mentioned that the cave house had in the distant past been a café and that buses used to bring visitors along the roadway above the terrace for refreshments and to enjoy the woodland views. Cave Cottage is on private property and permission from Apley Estates is needed to visit the area.

Cave Cottage, Apley.

Various graffiti and dates were noticeable on the walls. There was formerly an outside toilet a few metres away from the cave; this is now eroded and unused.

The gamekeeper mentioned that other caves exist on the Apley Terrace. There is a small one on the cliff near to Winscote cottages, a further cave lies just below and slightly right of Cave Cottage, further north on the terrace is a cave/bench feature and further still lie other caves. All of these features are on the Estate and permission must be granted before attempting access.

Eyton on Severn

Lee and I again set out into the Shropshire countryside in February 1998 to find the grotto at Acton Burnell. We were unsuccessful, so to save wasting further time we drove a short distance to Eyton on Severn. The OS map mentioned a cave near to the river. The cave turned out to be a manmade structure of three chambers and may well have been a cold storage area for meat as there are no windows to lighten up the rooms and only a single entrance door. There is also no form of fireplace.

The first chamber is circular with low headroom of 2m, the diameter being approximately 3m. The second room is in complete darkness and has rows of shelves and compartments carved out of the sandstone on three sides, resembling a form of larder store. A third chamber travelled deeper into the hill. A step down of at least 45cm is made in the floor level, but the room size could not be estimated due to the inky blackness. The room was not entered blindly in case it may have been an icehouse chamber.

Eyton on Severn.

From here we followed the sandstone ridge to a place called Eyton Rock where we were to find signs of former rock house-type structures. The first feature had been a circular room set into the ridge at an elevated height. A tree had grown into its roof and the front to the cave had fallen away downhill. The back of the cave had a bench carved into the sandstone with seven notches in situ for a wooden seat top. An arched window was carved into the left-hand side, which would have lit up the chamber. Pick marks were visible to the walls; an armrest was also carved into the benching beneath the window. Further along the cliff face was an area where notches for joists were noted and nearby was what I perceived to be a chimney flue. This had pick marks, showing that the slot had been hand cut. This slot then entered into a circular bore of 6in for approximately 2ft before coming out to surface in the turf layer on top of the escarpment.

The area of habitation, containing the window, may not last many more years due to the tree roots growing into the sandstone and heavy cracks have already appeared in the window and side wall.

It is noted in local history that Lord Herbert of Chirbury, an eighteenth-century philosopher, lived nearby in a large hall (now demolished). It may well be that these caves or cold storage cells were used in conjunction with Eyton Hall and Estates at that time, and now left to the elements and time.

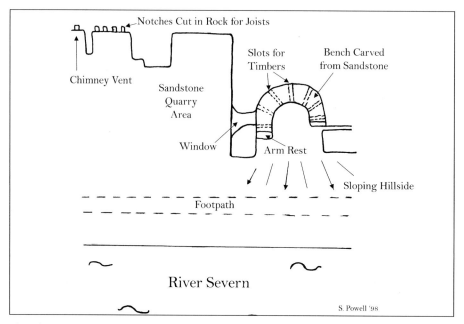

Plan of Eyton on Severn.

Eyton on Severn.

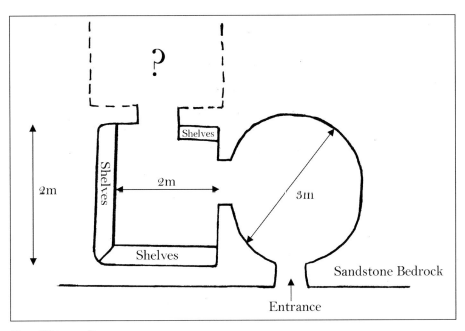

Plan of Eyton on Severn.

85

Gags Rock Cave

(27 September 1997
Revisited 18 November 1999)

The cave, visited by my son and me one autumn, is situated by the small village of Mose near Bridgnorth and is now totally obscured by a wooded copse used for breeding wild fowl. The cave, as remembered, is quite a remarkable site, as it is a totally natural cavern formed by the erosion of water, through fissures created by heavy faulting of the sandstone beds.

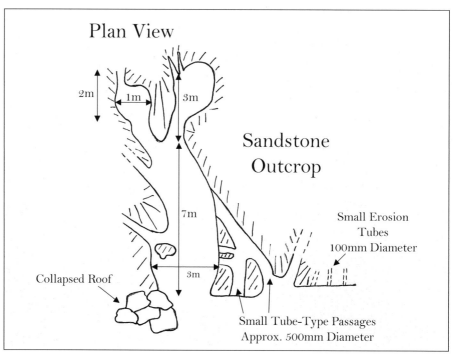

Gags Rock Cave plan.

Gags Rock Cave.

Interior of Gags Rock Cave.

The entrance is quite large, with smaller openings found to the right-hand side of the escarpment. All the passages follow a downward trajectory into the hillside; the smaller crawl-sized passages all meet into the main passage 4m from the entrance. At the rear of the cave (approximately 7m) the roof drops down to meet the floor, but three small passages can be viewed travelling further into the hill. Estimates of their length could not be determined due to heavy silting-up of the cave and insufficient light.

The sandstone cave is now situated some 88m above the level of the River Severn. Although the cave is heavily silted, it still retains the modest headroom of 1.5m.

A return trip was made to Gags Rock in which entry was made into a small chamber on the right-hand side and increasing the length of the cave by a further 3m. The left-hand passage appears to go further but is blocked to within 15cm of the roof by silt and rubbish (broken bottles, bits of iron, bricks, etc.).

A lady in a nearby cottage said that it was rumoured that the left-hand passage led down to the river and the right to other nearby venues. She said that slabs of rock had fallen from the roof of the cave, blocking off the passages. Whether it was my imagination playing games I am unsure, but the floor in the end chamber seemed very hollow.

Another tale is that a passage connects Gags Rock to Stanmore Hall, for a now unknown reason.

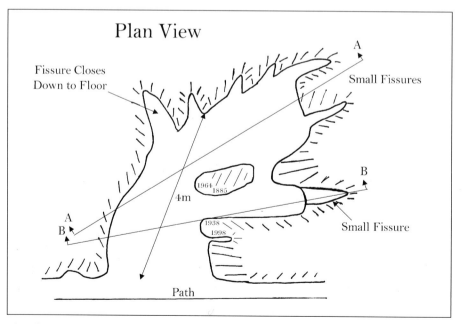

Plan of Sandybury Cave.

Sandybury Cave, Bridgnorth

The cave is found in a forested area towards the top of a valley, approximately 70m above the River Severn at Sandybury. The cave entrance would formerly have been approximately 3m high, but it is evident that erosion has collapsed the roof slightly so that the cave starts with a ceiling height of approximately 2m. The cave has been formed by erosion by water along faulted beds from the surface. The ceiling slopes to the natural dip of the sandstone beds, showing that the travel of water down one of the weaker beds has opened out the sandstone. The cave at its rear splits into two small pockets, which are connected by a small window-type opening and roof pillar of natural origins. Various graffiti cover the walls – some notable dates exist as 1885, 1938 and modern ones of 1964 and 1998. Small erosion fissures were noted beneath the cave lower down the escarpment.

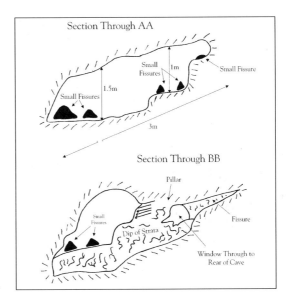

Sections of Sandybury Cave.

Sandybury Cave.

Bridgnorth Hermitage Caves

My first visit concluded that caves of various sizes were found on three terraces or levels, all in close configuration. The top terrace being the oldest contains the Saxon Hermitage (or chapel), as it was known later. This top terrace gives the best view of the Triassic sandstone strata, which is overlain by an unconformity of glacial drift termed 'The Cat Brain' formation. There are two further large caves on this top terrace beside the chapel, one of them reputedly the Witches Cavern. The caves are all stepped up from each other as the terrace travels uphill. There are many interesting features such as alcoves, ledges and beam holes and a flight of well-worn steps carved into the rock that now travel into fresh air. The room it once accessed disintegrated long ago. The chapel is also much in demise and featureless, compared to historical descriptions of it in the nineteenth century.

The second terrace is linked to the first and travels around the hillside slightly north-east to north-west as opposed to the top terrace north/south direction. The rock-cut rooms on this second section are the remains of the custodians' cottage. The rooms are interlinked by short passages, steps and extra openings made by vandals and nature; again small notches, shelves and alcoves are visible, carved into the sandstone.

The excavations on the lower terrace were small but numerous. Most of the dwellings here were constructed of three walls of sandstone and brick with the cliff face used as a back wall, into which small alcove-type caves and outhouse stores were excavated. The remains of roof flashings on the sandstone, forming a triangle shape, could determine the positioning of three pitched roofs. Stone corbels, which were seats for roof purlins, are also still in situ. This terrace also faces north-east/north-west. Old drawings indicate that possibly five houses were positioned here. An interesting feature to this level is the positioning of a cave at ground level. It is separated by a metre of bedrock with another cave directly above it that was once part of an upper storey to one of the rock houses. Another terrace exists below that is totally devoid of any tunnelling or building work and it is presumed that this was part of an area used for stone extraction, the stone most likely used in the construction of walls to the dwellings.

Old view of Bridgnorth Hermitage.

The Hermitage Caves have now deteriorated badly, large blocks of sandstone having fallen out of the roofs. Every part of them is open to the natural elements, which is helping to erode them, and vandals have added to their demise by sadly tunnelling through party walls and demolishing the remains of brick structures. During my visits from 1996 onwards the caves were visitable, but now, in the year 2000, they are fenced off with danger signs, indicating that they are close to imminent collapse.

Not much of the caves are visible from the roadside due to thick vegetation and great care is needed if the site is visited, due to cliff edges that are unprotected and also obviously from the collapse of roofing stone.

At the bottom of the Hermitage Hill, in the side of an embankment and covered in heavy scrub, were found two further caves. These were duly investigated. The first was 4ft high by 3ft deep and 3ft wide. The second was 6ft wide by 6ft high, but only 1ft 6in deep. Both caves may have originally been larger before erosion of the outcrop of sandstone, but one suspects that they were stores of some kind fitted with doors and frames.

The hermitage is of great antiquity and dates back at least to Saxon times. It is reputed that between AD912 and 924 Aethelward, the second son of King Edward the Elder and grandson of King Alfred, became the first hermit to seek refuge and sanctity.

It is not known whether the hermit's cell was as yet excavated into the sandstone but it can be established that rock-cut chambers were carved by the fourteenth century.

Aethelward is reputed to have been a learned scholar, interested in literary matters and ancient customs. His father died in the year AD924. We conjecture that Aethelward left the hermitage to succeed the throne due to his brother Edwin having died under mysterious circumstances some years previously. Tragedy was to occur again as Aethelward only reigned for sixteen days before dying and joining his father in a tomb at Winchester. His half-brother Athelstan then succeeded the throne.

John Leyland made a significant reference to these proceeding during the reign of Henry VIII when surveying England's antiquities. In 1539 his work brought him to Bridgnorth where he quoted 'In this forest or wood (Morfe) King Athelstan's brother led in a rocke for a tyme a hermiter life. The place is yet seen and called the Hermitage.'

There does not appear to be any reference to hermits until the rule of Edward III who brought hermits under royal patronage; throughout the

Hermitage Caves.

Staircase to upper chamber, Bridgnorth Hermitage.

fourteenth century, a royal patent and seal was required to become a hermit. This does not imply that the Hermitage was empty in the intervening period. The Revd R.W. Eyton, who produced the important book *Antiquities of Shropshire* left us with the following information from the fourteenth century pipe rolls.

In 1328 the cave chapel was the hermitage of Athelwildsten. In 1333 this changed to the hermitage of Adlaston near Brugenorth, by 1335 reference is made of a hermitage of Athelaxdeston. This is interpreted to mean 'The rock or stone of Ethelward'. The occupants were John Oxindon (1328), Andrew Corbrigg (1333), Edmund De La Mare (1335) and a Roger Burghton in 1346, who was presented to the hermitage above High Road, Brugenorth. It seems quite clear that the hermits did not reside at the cave long before moving on to other locations. The next literary tract comes from *Magna Britannia*, 1767, Vol.IV. It states 'Upon the brow of a hill of Morfe is an old cave, supposed to be the habitation of a hermit in which was a descent by steps into the earth to a great depth, but of what use it was is not known.' There was such another in one of the cellars of the castle and was thought to have a communication with the other under the river.

This began a phenomenon of secret connections to various parts of Bridgnorth, irrespective of whether the River Severn stood in its way.

There is mention of a tunnel linking the hermitage with the friary in Low Town.

Another connection is the hermitage to Hoards Park Ancient Mansion. There were extensive vaults, cellars and passages beneath the old house, some of

Hermitage Caves.

which were reputed by Hubert Smith in 1877 to have been bricked up. Similar vaults and passages were noted at the friary. It is quite easy to see how tunnels from one place get integrated with those of another place and then the fairytale tunnels unfold. Cann Hall, which stood at the bottom of Hermitage Hill, also became entwined into the legendary hermit's tunnel.

Hubert Smith, Town Clerk of Bridgnorth, made the best-known story/report in 1877. Working on behalf of the Shropshire Archaeological and Natural History Society, he first enquired to the Hermitage caves custodian's wife about the subterranean passage. She told him grand tales and hearsay stories about a rumoured chest of buried treasure, presumably deposited by a witch who had lived in the cave next door to the hermitage chapel. Smith mentions that his faith in the stories was very weak and yet he knew that the traditions had a respectability of age. Hubert also enquired of the custodian's wife about whether she remembered a couple of boys digging at one corner of the hermit's cave and coming to some steps. 'Oh yes sir', she replied, 'and the bailiff of the farm, passing by, was very angry and made them fill it up.' 'The right hand corner of the cave?' Hubert asked. 'That's it.' And this last was said with such emphasis as to leave no doubt of her firm belief in the existence of the passage.

Hermitage Caves.

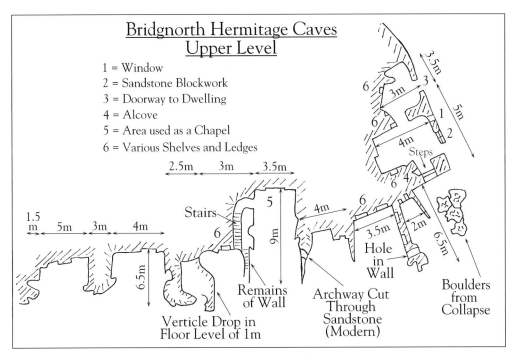

Bridgnorth Hermitage Caves
Upper Level

1 = Window
2 = Sandstone Blockwork
3 = Doorway to Dwelling
4 = Alcove
5 = Area used as a Chapel
6 = Various Shelves and Ledges

2.5m 3m 3.5m

1.5m 5m 3m 4m

3.5m 3m 3 5m 1 2 Steps 4m
6 6 9m 4 6
Stairs 5 4m 6
6 3.5m 2m 6.5m
Hole in Wall

Remains of Wall
Verticle Drop in Floor Level of 1m
Archway Cut Through Sandstone (Modern)
Boulders from Collapse

6.5m

Plan of Hermitage Caves.

Her interest was considerably increased when Hubert unfolded his project for examining fully as to the existence of an underground way; especially when she found out he had full authority to do so. The custodian was employed on the project, with some direction given by the bailiff of Hermitage Farm as to the cave's position. All the soil and rubble was cleared from the cave, right down to solid bedrock, with the exception of a small portion of comparatively recent brickwork. No entrance was ever found. Hubert Smith had a professional plan drawn up of the hermitage chapel, which pointlessly marks the spot of the entrance to the elusive tunnel.

It is said that three years later a Mr Seamer, an occupant of an adjoining cave, excavated the site and he too found nothing conclusive. It is known that rewards have been offered in the past for definite information about the elusive tunnel. In 1963 the author Sheena Porter wrote a children's novel entitled *Jacobs Ladder* which was based at Bridgnorth and the finding of a secret tunnel which ran from one side of the River Severn to the other, then mysteriously climbed by a series of subterranean steps to a hidden crypt. The crypt, however, is on the high town side of Bridgnorth and not the Hermitage Hill side. An amusing book.

Old engraving of Bridgnorth.

The Hermitage is said to have been supplied with water from a nearby spring. This has been described in 1877 as seldom used and seems to issue from beneath the roots of a hawthorn tree, collecting in a shaded cavity 5ft 6in long by 2ft wide.

Hubert Smith gave an interesting account of the Hermitage in 1877. It is entered through a small door with rock on either side which, with the addition of a few blocks of sandstone, closes in the front to some height. The first and largest cave is now roofless, while on the left is a small side cave, called on the plan the 'Lower Cave' which is part roofless. The roof of the entrance cave has crumbled away to the first arch of the interior cave chapel. The oratory or chapel is still intact, with its rudely-arched roof in now red sandstone. It will be seen on the plan that the oratory or chapel is smaller than the front cave, but it is the most interesting and complete section of the ancient Hermitage. A brick pavement has been, at some recent period, laid down on its floor of solid rock, while on the left of the chapel a lofty passage in the rock leads up a flight of steps cut out of the rock to the upper cave. As you enter the upper cave there is an opening or window through which you look down into the front cave below. The small upper cave is now roofless and partly floorless. Such is the Bridgnorth Hermitage, which appears to have consisted of four rock caves or chambers, one being used as a chapel.

Hubert Smith mentions that the family in the custodian's cottage also had the occupation of the Hermitage and its precincts. He found them a worthy couple of the labouring class, who had as much attachment for their rock abode as if it had been a palace. It was an excellent specimen of a cave dwelling, many of which formerly existed about Bridgnorth. The hand of man had done little to

Old engraving of Bridgnorth.

COTTAGE in the ROCK, BRIDGNORTH.

render it habitable; a chimney and fireplace had been built up at one end and a doorway and window had been fitted into the rock façade, all had been accomplished with very little aid of masonry. Smith mentions that on his first visit he found the good wife seated at her little round table drinking a cup of tea. An old clock on the right stood near some broad wooden steps leading up into a narrow cave bedroom, which had a curtain before the entrance instead of a door. On the left of the fireplace, another cave served as a sort of storeroom. The kitchen and bedroom were narrow and not very lofty. The good wife idolised all that belonged to the Hermitage; her husband and herself had lived in the cave cottages for the last fourteen years. Smith also believed that it would break the couple's hearts if they were to have to leave their humble retreat.

There was a sketch drawn by Powell, the artist, depicting brick–built cave cottages and their occupants in the early years of 1828. The Hermitage rock houses were drawn again in 1854 by the artist Brooke, showing No.5 dwelling.

The caves were reported in 1883 by Charlotte Burn in *Shropshire Folklore*. She mentions a smaller cave known as the 'Witch's Cave' which no longer existed. It was said that the witch who lived there used spells to bring horses and carts to a stop on the road running past the caves. However, the horses probably

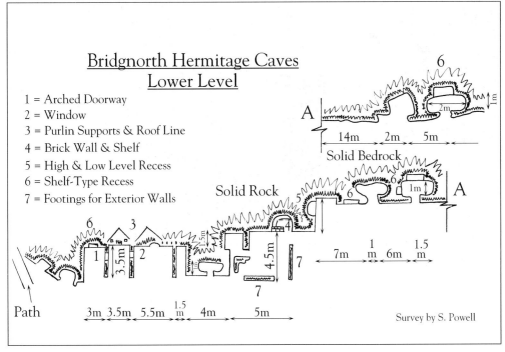

Plan of caves, Bridgnorth Hermitage.

Above and opposite: *Caves, Bridgnorth Hermitage.*

stopped from fatigue after pulling their loads up the steep hill. Charlotte Burn also mentions the supposed secret tunnel, and that a custodian in 1881 had also dug over the floor of the hermitage and had found nothing.

On page 85 of G. Jackson's *Shropshire Folklore*, published in 1883, she states that in Bridgnorth there are some caves in the side of one of the ruddy sandstone cliffs, overlooking the Severn. Once used (one of them still used) for human habitation, one of these now empty and deserted caves is known as the hermitage. Here the rock has been smoothed and shaped into rounded arches and there are evident traces of a little oratory or chapel.

In 1899 H. Thornhill Timmins wrote his book *Nooks & Corners of Shropshire*. He too found the time to visit the Hermitage. He states:

> *Time and neglect have played sad havoc with these singular grottoes, but their main features are still discernible. The chapel, an oblong chamber hewn in the living rock, is now partially open to the sky, though the chancel, with its rudimentary rounded arch, remains intact and there is a shallow round-topped recess in the eastern wall, where the reredos usually stands.*

Alongside the chapel we find the hermitage proper, a low dark cell communicating with the chapel by a small aperture, now blocked by the large ungainly brick oven which defaces the interior of the chapel. There is an apocryphal tale that a passage formerly existed, connecting this hermitage with Bridgnorth Castle and that chests full of priceless treasure lay hidden away somewhere amidst the recesses of the rocks. But, needless to say, no treasure-trove has ever been brought to light. A few paces distant stands a lowly cottage dwelling which is excavated, like its neighbours, from the solid rock. It was until recently tenanted by a family of modern troglodytes and is still used in the daytime by the good woman who has charge of the Hermitage, so let us glance within as we pass.

Upon entering, we find ourselves in the living room, whose roof, walls and floors consist of the native sandstone, a warm weatherproof covering, though blotched and variegated with many a mottled stain. A short stepladder gives access to a small upper chamber, with seats roughly cut in its rocky walls and a window pierced through the outer one. The interior was drawn beautifully by Timmins, showing all the domestic equivalence to that of ordinary houses. The lady was without doubt one of the Taylor clan. Five generations of the Taylor family lived as custodians to the caves. Cyril Taylor's grandmother lived there from 1898 to 1911, until part of the family moved on to live in Birmingham. His mother's family had lived there for many years and Cyril Taylor could recall living there until 1928, the rock house then being retained as a holiday home

Caves, Bridgnorth Hermitage.

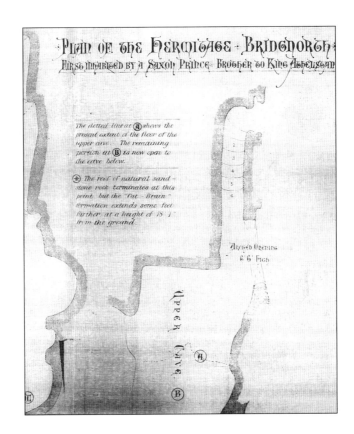

Old plan of caves,
Bridgnorth Hermitage.

until 1930. The custodian's duties, besides looking after the chapel, was to patrol the adjacent woodlands to cut down on the poaching of game.

Rotha Mary Clay's *Hermits and Anchorites of England*, published by Methuen in 1914, mentions the caves and contains a photograph showing the chapel with door and windows. Another old brown and white photograph exists in the Shropshire archives showing the inner chapel cave with carved ribbed arches and alcoves. The caves were very popular as a tourist attraction between 1900 and 1915 and are featured in the official town guide of Bridgnorth of 1907. Local papers and magazines have done many features ever since. The *Bridgnorth Journal* of 31 December 1999 produced a short article entitled 'Hermitage Forever Shrouded in Mystery' for a special millennium edition of their paper, recalling many of the historical places in and around Bridgnorth.

It is said that Cyril led a comfortable life while living in the rock houses, living off the land until November 1928, when it is reputed a pheasant was found in one of his rabbit traps. This, in turn, led to his eviction and hence only the temporary use of the caves from 1928 to 1930, as his holiday home. Whether his use of the caves in this latter period was legal we can only guess.

Bridgnorth

The underground delvings in and around the town have for many years been a curious attraction to both tourists and historians alike. The Tourist Guide to Bridgnorth for the year 1875 gives us such an insight.

> *If one approaches the town from the river and follows one of the winding roads, which diverge left and right from the River Bridge, he will be struck by the sight of human habitations. whose only approach to ordinary dwellings consists of a facing of brickwork erected against the external surface of the rock out of which they are hewn.*

So extensive were the excavations into the sandstone bedrock that in 1934 it was stated that in the East Cliff of the town the total number of caves was around 250. H.L. Cunnington, a member of Bridgnorth Historical Society, reported that it would be important for one or two studies to be made of them, but it appears that this did not happen. My own study on the town has sadly produced far less numbers.

As with any study, we need a starting point, so one sunny weekend I was accompanied by my family to explore the caves. We started at one place called Underhill Street, an area well known for its rock dwellings and excavations and also for the reasons of accessible parking. An early manuscript document of 1398 from the rentals of St Leonard's Chantries stated: 'From John Underhill for the cave and garden rent due at the feast of the finding of the true cross – fifteen pence.' This shows us that the excavations began a long time ago and it also shows the importance of Mr Underhill in that his name was adopted for the street. It is thought that the cave, now known as Lavingstone's Hole, was the one formally owned by Mr Underhill, but much altered as for reasons explained later. Underhill Street is now a pretty-looking landscaped garden with interlinking paths, the caves and gardens are lit by coloured spotlights by night. The first chamber visited is termed Cave No.1 and is of large dimensions, approximately 7m wide by 11m long, it also has a lofty ceiling height of at least 3m. The chamber, due to its size, required a roofing pillar to be left in and this is found slightly off-centre. Due to gradual erosion of the roofing strata on the

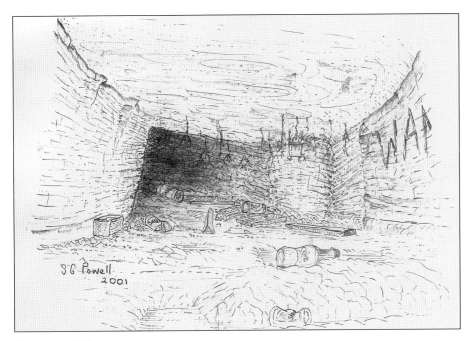

Lavingstone's Hole.

junction of the pillar (the sandstone beds are tilted at approximately thirty degrees), a second support pillar was constructed to help stop further slippage, flaking and upward migration of the rock strata. (This would have eventually lead to the cave collapsing.)

This support pillar is made of large sandstone blocks. There are many pick marks left in situ, particularly around the roofing pillar, and these date back to the cave's original excavation. The main doorway has been improved with brickwork and is found to the left of the chamber. It was noted that a further partially hidden doorway and passage exists to the far right of the cave. A small portal window lets in a small amount of light and there are various beam holes in the bedrock where there were once storage shelves.

A small storage cave is found next to the doorway to the right and is marked cave 1A on the survey. Cave No.2 had been blocked off with sandstone block masonry, possibly to support the cliff front and cave structure, as the natural bedding of the rock strata is faulted close by. Some of the masonry had been removed from the wall and this allowed me to peer into the chamber, which had approximate dimensions of 3m in width by 2.5m deep. This cave also shows a lofty ceiling height of approximately 3m. A large galvanised tank had been fitted into the cave at some point in time but its purpose is unknown.

Cave No.3 has one large aperture that spans some 5m in width by 2.5m high. The chamber that is directly in front is only 2m deep. In the corner of the cave there is a small shelf and holes for timber poles used for further shelving. Adjacent to this chamber are two smaller caves termed 3A and 3B, obviously small stores. There is a further opening marked as 3C, this is of entirely natural origins and is an eroded fissure. Cave No.3 would have been the back section of a cave house, the remainder being made up of brickwork.

Cave No.4 consists of a large chamber approximately 5m by 4m by 2.5m high from floor to ceiling. It is divided at its entrance by a small pillar. Unfortunately the cave was not entered due to the presence of old mattresses and bedding duvets. The place is obviously, at the time of writing, a residence for some unfortunate soul. To the left of the cave is a small excavation, 4B, part covered by a section of brickwork; this is likely to be a small store.

Area 5. Although there are no cave-type excavations, it is interesting as it shows the purlin holes and rooflines where dwellings were built up against the sandstone cliff using the cliff as the back wall.

Cave No.6 is one of Bridgnorth's more famous spots. The cave, known as Lavingstone's Hole, consists of a large entrance chamber approximately 3.5m wide by 6m in depth and has a very high ceiling line of at least 6m. The cave is

Underhill Street Caves.

*Underhill Street
Caves.*

cut into the rock at approximately forty-five degrees to the cliff face. A small access tunnel is found in the roofline, this also continues into the hillside for a further 20m. The tunnel inclines to the same pitch as the natural bedding planes and cambers gently uphill. It is of only small bore, 1.4m in width and between 1.1m to 0.6m in height, and various pick marks are evident throughout, showing that no explosives were used in its forging. My own good fortune was being able to access the high level passage by a long length of slatted fencing placed into position as a sort of ladder by persons unknown. A surprising aspect on climbing into the tunnel was finding it full of rubbish, some thrown into it from the chamber below. The tunnel was also covered in graffiti, one of the dates carved into the rock is from 1746, and this is found at the far end of the tunnel.

The cave and tunnel are named after the English Parliamentary Forces Engineer Colonel Lavingstone, who in 1646 was part of an army detachment sent to besiege the royalist forces occupying Bridgnorth Castle. After failing to breach the castle walls, Colonel Lavingstone proposed the boring of a tunnel to reach beneath the royalist powder store, fill the tunnel with gunpowder and ignite it – hopefully blowing up the enemy ammunition. Before the excavations had reached their destination, the castle was unexpectedly surrendered and the tunnel was abandoned. On the last shift the workmen had cut their picks into the rock to a depth of 100mm in either side of the rock face preparing the removal of a large slice of stone. The cave and tunnel were explored by Bridgnorth Historical Society in 1934 and it was explored and measured again in 1943 by members of a nearby RAF base. One of the group, a mining engineer known as Cadet Stevens, stated that the tunnel was pierced by experienced workmen, his survey quotes:

> *Axis of tunnel points 240 degrees (time)*
> *Length of tunnel 70 1/2 feet S 59 degrees W.*
> *Mean width of floor 6 Ceiling 5*
> *Height of S.E. side of tunnel 3'3"*
> *Height of N.W. side of tunnel 2'9"*
> *Rise of floor 3'3" (Approximately in total length of tunnel)*

It is said that due to the destruction from the Civil War in 1646, many people from Bridgnorth began to move into caves that already existed and to excavate fresh ones.

Lavingstone's Hole became a focal point for tourists in 1934. A pathway led directly to the cave, which at that time was partly boarded over as it was to be converted it into an outhouse. The cave had been considerably cleaned out in

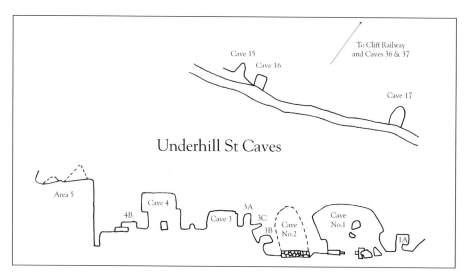

Plan of Underhill Street Caves.

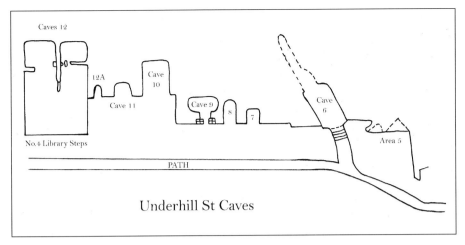

Plan of Underhill Street Caves.

preparation for Mr Maiden, the landowner, to construct a staircase and platform. This reached most of the way up towards the tunnel. A short ladder was fixed from the platform to access the mine tunnel. It was said that lady visitors need fear not the slightest embarrassment by climbing the short ladder. Mr Maiden charged a nominal fee to persons inspecting it, but assured that Bridgnorth Historical Society members would be admitted free. Lavingstone's Hole became the centre of attention again in 1997 when a party of schoolboys climbed into the tunnel but became petrified when it was time to climb down. The emergency services were called out to rescue them. Since spring of 1999 the caves on Underhill Street and Low Town have been sealed off from the public with galvanised steel fence panels.

Following the path uphill towards Library Steps, we pass caves No.7 and No.8. These are small and choked with earth and rubble. Cave No.9 is part covered with ivy and has a brick-vaulted structure for its portal. A ceiling of natural sandstone could be viewed inside; this small cave was also semi-blocked by rubbish and rubble. Further uphill caves No.10 and No.11 can be viewed, but not visited as they are positioned in part of a private garden. These have large brick-vaulted entrances. These caves formally belonged to the dwellings of No.2 and No.3 Library Steps, which were demolished at some time past.

No.4 Library Steps or April Cottage is host to caves 12A and 12. The cottage is built into the rock face and two rooms have been excavated. The deeds to the house describe it as a cave house. The house owners kindly sanctioned a visit for Shropshire Caving and Mining Club in 1998 from which the information was gathered.

To the left of Library Steps a rough track through brambles and nettles led to caves No.13 and 14. The first was probably used as a store and has dimensions of 2m wide, 2m in depth and 2m in height. Cave No.14 is double the size but holds a 2m ceiling height. Two recesses for shelving or cupboards have been carved into the cave walls, showing it to be a habitable room at some time past. The brick frontage to this former dwelling has long since slid down the escarpment. The only proof that it was a cave house are the marks of a roofline indented onto the sandstone cliff. The cave shows vestiges that a tramp occasionally uses this spot for a habitation.

Following a circular walk, we head for St Mary's Steps. Walking downhill on the left-hand side we spot what are caves No.15 and No.16. These are on private property and they have both been backfilled with rubble and garden cuttings. Cave No.17 was visible towards the bottom of St Mary's Steps; this was probably a small storeroom.

Doubling back on ourselves we find, at a place known as New Road Corner, caves No.18 to 28. There are eight numbered brick-arched openings, which are blocked off by timber slats. Due to the amount of overgrowth at the time of my visit I was unable to determine the size of the caves' interior, but surmise some to be small stores to the houses that occupied the site before the modern road-widening scheme. Two further entrances were located, but found to be bricked up. Above one of the caves are the interesting features of the roof pitches and beam holes to a former attached structure. There were also signs nailed across

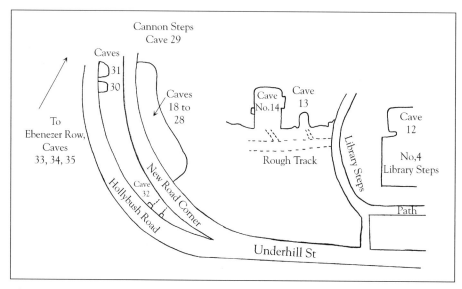

Plan of Underhill Street Caves.

the cave entrances: 'Keep Out – Falling Rocks'. New Road area is represented in 1825 by a sketch drawn by artist J. Powell. The sketch shows many rock cut openings and chimney breasts and flues protruding through the bedrock.

A further painting was done by another artist at the later date of 1829, showing a close-up view of two cave cottages. A cobbled courtyard is shown surrounding the doorways and some of the inhabitants are portrayed relaxing in fine weather.

Just uphill from New Road is a place called Cannon Steps and to one side of the steps are possible features of a former cave (No.29). It is recorded that a cellar existed beneath the tower of Bridgnorth Castle and speculation has it that a secret tunnel leads from here to the hillside at New Road. The tunnel portal was reputedly seen covered over by the brick walls used to construct and retain the castle walk. Below New Road is Hollybush Road, there are rock cut features here, namely two brick-vaulted entrances (Caves No.30 and 31). These lead to chambers now used for garages by nearby house owners. Cave No.32 is found in the brick retaining wall near to the junction of Hollybush Road and New Road, and it is quite a puzzle; the bricked-up portal of a tunnel is clearly visible and the tunnel or void must surely travel beneath the road into the sandstone escarpment. For what purpose is anybody's guess.

Moving on to Ebenezer Row we find a row of old cottages in the cliff, behind them are three small caves.

In St Leonards Close there is a seventeenth-century house known as the granary and in its grounds was once a secret tunnel that led to the nearby thirteenth century sandstone church. Yet again we have a potential tunnel, with nothing to prove its existence or if it ever existed other than in peoples imaginations.

By the 'Fernicular Railway' cliff are caves No.36 and 37, these are visible on the left-hand-side of the escarpment. Both have brick-vaulted frontages but cannot be visited due to their inaccessible position on the railway grounds. A number of other caves are reputed to have been destroyed by the building of the Fernicular Railway in 1892. This has been mentioned in the Bridgnorth Town Guide and *History of the Castle Hill Railway* by C.F. Gwilt. It states that the work was much hindered by the discovery of caves, which were broken into at the lower end of the incline; a cave so large being found as to necessitate the building in of massive masonry columns and strong wrought iron girders for support.

In the later part of the eighteenth century a potent local brew known as 'Eve's Beer', later known as 'Cave Beer', was produced and stored at a cave in Underhill Street. This is probably the property addressed as No.6 Underhill Street, and known as The Malt House. The historian Cornes wrote:

New Road Corner.

There is a cave on the south side of the passage (stone way steps) which was remarkable for being the repository of that excellent Cave Beer, contained in a large vessel or tun holding more than five hogs heads.

He gave the dimensions of the cave as 33ft in length by 17ft in breadth.

At a later date messrs. Ind Coope's and Messrs Davies and Locket were recorded as having very large caves to the rear of their premises. It is hinted that the cave originally used for Eve's Beer was taken over and used by one of the latter for storage of their own products.

The Victoria County History of 1908 also makes reference to a cave of corresponding measurements that held five hogsheads of 320 gallons of beer. The Malt House caves in modern times are fronted by a double-arched brick entrance; the left-hand doorway leads to a large cave 12m long and 4.5m wide. The roof is barrel vaulted with narrow brick-ribbed arches springing from the side walls. The floor has been laid with bricks and to the rear of the cave are five brick bins. The right-hand doorway leads into a chamber 4m wide by 3.4m deep. The two caves are interlinked in the far left of the second chamber by a short passage.

To the left of the first entrance arch inside the large chamber, a passage can be seen to travel to a network of further caves consisting of three chambers. No.5 Underhill Street has two caves which are entered through segmental brick-arched openings and a passage of 2m in length, beyond this is a huge cavern 5.5m wide by 7m long. The ceiling is slightly domed and the walls are straight and well cut. The floor to this cave is of brick with a small area of cobbles, an interesting feature set back into a recess are two brick-vaulted storage ledges divided by a single skin of brickwork, probably used for storing beer kegs.

The second cave is made up of two rooms. The first is of an irregular nature and a roofing pillar of solid bedrock has been left in for support. The second chamber is rectangular and near to the doorway is a small recess cupboard. The first chamber is approximately 5m by 5m of which the pillar takes up 2m square. The further chamber is approximately 3.5m long by 3m wide. The caves are unfortunately backfilled with rubble and are only just accessible for surveying.

The last available cave at Underhill Street is found next door at No.4; the premises at present house a commercial shop. However, set into the cliff face at the rear of a large yard is a cave 4m wide by 4.5m long. The ceiling is slightly domed and is 3m at the highest point. The entrance to this cave is brick-vaulted to its full width so as to leave ample room for the parking of a vehicle. A small hole in the left-hand wall connects into what is described as a semicircular rock-cut oven, with a brick laid floor. The floor and back wall of this feature show burn marks.

Drawing of Friary.

It is possible that the structure was used in conjunction with the brewery trade as a malting oven. It is on record that in 1885 Nos 4, 5, and 6 Underhill Street were used for commercial purposes connected to the brewery trade.

Although not much publicised, Bridgnorth was heavily involved in smuggling activities. Lines of merchandise were shipped up the River Severn and were subsequently dispersed through many of the town's taverns and shops. It is rumoured that some of the subterranean chambers of the town were used as holding points. Similar practices were in operation at Worcester, further down the Severn.

The 'Magpie Inn', formerly the 'Bee Hive', was said to have a secret tunnel running from the stables in Cartway down to the river.

At No.92 Cartway, two arched recesses are visible, these are the remains of a rock house. Closer inspection shows that the two recesses are really one chamber 4.6m wide by 3m deep by 2m high, but are partitioned off with a single skin of red brick. A small cupboard niche is carved into the back wall.

Close by, at No.95 Cartway, there is a small cave used as a storeroom and further down Cartway is a cave used as a garage. There are two other cave houses in existence on Cartway, complete with the original brick frontages, unfortunately the doorways and windows have been blocked off and above the doorways a plaque reads 'This cave was occupied as a dwelling until the year 1856.' The address of one of the rock houses is known as 78 Cartway. Other caves have been noted to exist at the rear of No.44 Cartway.

An area of Low Town, known as Riverside, was visited next. In a cliff face covered with ivy, brambles and overgrowth, were found a series of small storage caves. The first, a double recess, had for support a small roofing pillar, the second was a singular cave with part brick pillars to its entrance. The third cave was identical but without brickwork. A fourth cave, with an extended chamber to its left also lay hidden behind the veil of shrubbery. Walking along Riverside, back towards the town, a series of four distinctive large caves becomes visible. One is used for a wood store, another for a garage, the third could be viewed through a window-type entrance (the main doorway being blocked off with masonry). The cave inside appeared to have a depth of 5m and a width of corresponding dimensions. It was noted that it was part backfilled with rubble. There was a noticeboard above the window stating 'No campers'. This was obviously meant to deter youngsters from congregating in or around the cave. The fourth cave was completely isolated and shut off by a brick wall.

The cave used for a wood store was approximately 6m in height and may have originally been a boathouse due to its proximity to the river. Slightly north of this group of caves is a small store cave, 2.5m wide by 1.4m deep and 1.4m high.

Friars Street Caves.

Closer still to Cartway is a large car park that backs onto the cliff base. Cave remains can be seen in the cliff wall and there is a distinctive bricked-up archway that I conjecture leads to a subterranean chamber. The house next to the car park, No.30 Riverside, has a doorway in the cliff that leads to a whitewashed chamber with a narrow barrel-vaulted roof, while narrow-ribbed brick arches supplement its construction. It is approximately 3m deep by 2.8m wide with a 2m ceiling height. A similar cave exists at No.6 Riverside. The most interesting caves on Riverside are found at house No.29, which is situated just below Bank Steps. There are two caves situated in the cliff, which are in close proximity to the existing dwellings. The first is 10.6m long by 3.7m wide. A brick fireplace is positioned in the left-hand wall, of which a small portion of the chimney is exposed to the external elements. There are three low recesses carved out of the bedrock on the right-hand wall with a further recess on the back wall, these vary in height from between half a metre to a metre. The ceiling is a standard 2m high. The second cave is also 10m in length but much wider, approximately 6m. This cave extends beneath Bank Steps and to support its span a roofing pillar has been left in situ. The cave, at a later date, was divided into four by single brick partitions. The compartments are presently blocked by rubble, but it is possible to see the bricked-up doorway that led onto Riverside. It is presumed that these caves pre-date the dwellings in front of them and may possibly have been used as warehouse facilities.

Another curiosity of Bridgnorth is the Infirmary Hospital that was situated between Listly Street and Hollybush Road. This too had its underground delvings, and these were reported by the Bishop of Hereford in 1890 to a meeting at the Town Hall. He said that:

> A cleaner place he was never in, but certain defects were prominently imprinted upon his mind. He remarked that the mortuary and operating theatre were not suitable for their intended purposes as they were in caves carved out of sandstone positioned at the back of the hospital.

The situation was rectified, however, as a new infirmary was opened six years later.

Bridgnorth Friary was founded sometime in the mid-1200s and was extended and much altered over coming centuries. It was eventually demolished in 1861 to make way for Southhalls Carpet Factory. Towards the end of its lifespan, Cox's *Britannia* 1720 states:

> In the courtyard there are vaults underground which run parallel to the house for some space and extend themselves several ways, but how far in some places is unknown.

Friars Street Caves.

The friary is mentioned in the *County History of Shropshire* in which reference is made to *Magna Britannia*, pages 693-94, in which the friary was said still to retain plain marks of its ancient magnificence and some subterranean structures vaulted in stone.

This was to be proven in more modern times as the Shropshire Caving and Mining Club mentioned in their 1989 winter newsletter *Old Tunnels* that excavation work at the Bridgnorth Friary had found a network of tunnels. The club, however, never mentions whether they looked at them or if they had been

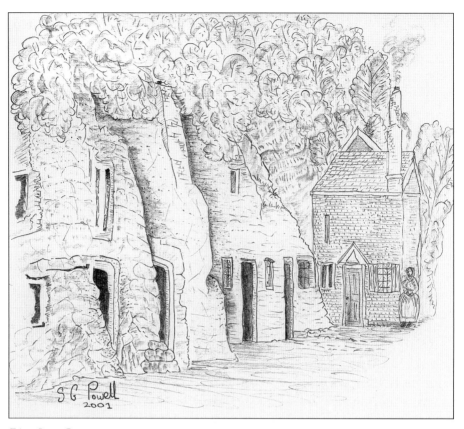

Friars Street Caves.

surveyed. Birmingham University Field Archaeology Group did, on the other hand, as part of a project, spend several weeks excavating the site prior to the building of the new housing estate (now known as The Friary). They uncovered the end of one of these passages, which resembled the hearth of a chimney with seats either side of it. The height of the cavity was such that a person could walk into it upright. It was walled on both sides and arched in stone. The secret passage that tradition says connected the friary to the hermitage caves was not found. However, a culvert-type tunnel was uncovered from the friary to the Severn, obviously used for foul water drainage, and it is possible that this was turned into folklore as a long forgotten secret passage, as is the case with many ancient and old buildings.

The caves and rock dwellings at Riverside and nearby Friars Lane were recorded in time by the artist J. Powell during the years 1825–30. Most of the dwellings visible in the drawings are completely carved out of the sandstone escarpment. The only brickwork visible is that of the chimney flues. There is in modern times the remains of a three-chambered cave system at 54 Friars Street. The main cave is 5m by 3m by 2m high and there is a lower recess of 2.7m adding to the depth of the room. A doorway to the left gives entry to the second chamber of slightly smaller dimensions, 4m by 3.5m. There is a redbrick outer wall with a window-shaped opening, this chamber also has an immaculate brick fireplace in situ that is capped with a wooden mantelpiece. A chimneybreast carved from the rock takes up the space between the mantel and ceiling. Moving left, we find the last chamber, this is only 2m by 3m and is unfortunately full of debris. At the site of the former cottages of 1-4 Friary Street there is a small blocked cave.

A further cave has been turned into an indoor garage by its owner. At the terminus of Friary Street there is a passage known as Granary Steps. There are two caves delved into the hillside beneath them. In the vicinity of No.3 Granary Steps, an entrance is found to a chamber 2.8m by 2.0m and a smallish ceiling height of 1.7m. A smaller chamber extends back 2.2m by 1m wide. On the back wall of the cave is a blocked-up doorway. The other cave is entered by a brick-arched entrance. The chamber is oval and 5.5m long by 3.7m wide. A further brick arch that led to a garden is now blocked up.

In and around Bridgnorth the collapse of ground has helped in accrediting the myth of the town's unseen subterranea. One such passageway was said to have been blocked up at Hoards Park Hall to stop straying sheep from entering, another story claims a schoolboy was asphyxiated by bad air in a tunnel here. However, a factor that has helped spread the tales of Hoards Park was the siting of an old culvert drain nearby, which magnified in proportion from storyteller to storyteller.

Sometime at the end of the nineteenth century a heavy old-type steamroller caused the road to collapse at Postern Gate. A local tradesman recalled how he had stepped down into an underground passage. Similar tunnels are reiterated to have been seen and entered in Poind Street and High Street.

It was stated during 1943 that local builders Messrs Meredith had stumbled on an old tunnel near to High Street. Although references report these phenomena, there is never any concluding evidence of how far these tunnels travel or in which direction and one can only assume that the finders of such tunnels never explore them. Other subsidences occurred in the lawn behind The Hollies in St Mary's Street and another in Crown Yard.

A local Bridgnorth historian, Mr. E.H. Pee, mentions that these subsidences were probably connected to an old sewer system that fell into disuse in 1855. The entrance to one of these large drains was once to be seen in the cellar of a house at the bottom of St Mary's Street. Unfortunately the house was destroyed by a bomb blast which also demolished the drain and cellar structure.

The only known tunnel of sizeable dimensions that can be walked through is the old railway tunnel that was first constructed in 1860. It runs straight through the sandstone hill from Railway Street and appears to daylight at Friars Street. It has been abandoned since 1963. There is much water pouring into the tunnel through its brick lining due to the nature of the sandstone above and a large pool has formed on the track bed. The engineer who built the tunnel was a Mr Dowell. There was only one reported incident during its construction and this concerned a cave-in. The incident occurred at 3 to 4 o'clock in the early hours and fortunately few men were working. Those who were, finished their nightshift unscathed. The tunnel was formally opened on 6 October 1860 and a horse-drawn coach, full of civic dignitaries, was driven through its entirety. It is stated that the mayor handed over five sovereigns from his own pocket to pay for the entertainment of the tunnel navvies. This was spent on a lavish banquet for sixty persons, hosted by Mr Dowell and held in the tunnel, with the catering supplied by the Ball Inn. A press report states:

> *A most pleasant evening was spent toasting sentiment with song following in quick succession, and when the health of the mayor was given, the bociferous applause and deafening cheers ran through the subterraneous passage, showing how they appreciated the mayors' munificence.*

The tunnel was formerly owned by British Rail, who paid for its structural upkeep. It was once suggested using the tunnel as a municipal fallout shelter back in the cold war era of 1981 but the idea was never taken seriously.

Other caves in the Bridgnorth vicinity include Dracup's Cave; this is named after the artist Anthony Dracup, who embellished a chamber with columns, arches and a vaulted ceiling until it resembled that of a church interior. It is located at SJ720 930.

On the main road between High Rock and Lower Lodge, there are caves incorporated into the cliffs of 'Jacob's Ladder' and Pendlestone Rock. And, situated on the B4364 Ludlow road we find 'Ye Olde Punch Bowl Inn'. This dates back some 500 years, becoming a tavern in 1785. There is a secret tunnel, which travels from the inn beneath the road to a further cellar, perhaps related to smuggling activities?

On the opposite side of the River Severn and close to Bridgnorth is the small village of Quatford. There is a row of cottages facing the main road and in the rear gardens are two cave entrances. The first is of large dimensions, shaped like an inverted 'V' and it appears much enlarged from erosion and spalling of the surrounding sandstone. The second cave entrance is of a much smaller bore and is decorated with a mock spider and web which is approximately 1m in diameter.

New Road Corner.

The nearby lane that links Quatford to Sandybury also displays some form of cave house burrowings, the remains of which are two brick support pillars and a hollowed out small chamber with a side connection that gives a hint that it may have been a chimney vent. To the front of these features is a drainage manhole carved from the living rock. The remains are all very sketchy but represent that of a possible former dwelling. The area in question looks to be in the early stages of preparation for future development.

It is mentioned that a former curate of Quatford, John Higgs, served the parish for sixty-eight years, passing away in 1763 aged eighty-eight. He was a clever man, versed in Latin, farming and building skills.

It is said that he hollowed out caves for use as a stable and pigsty and later on he carved a staircase in the rock to which at the base was a small gallery with seats. This became known as the Well Chamber. John would spend many hours here reading or deep in thought at life's issues.

It is unclear if these caves exist or whether the two caves mentioned earlier have any connection to John Higgs.

Hawkestone Park

There are approximately eleven caves and tunnels to be found at Hawkestone, most of which were excavated during the eighteenth century for the landowner Sir Rowland Hill. The purpose of these features was to form an exciting adventure within his landscaped gardens.

The subterranean features are found on three promontory points. Red Castle Hill, Elysian Hill and a long sandstone scarp known as the terrace, part of which incorporates Grotto Hill. Some of the caves are really overhangs or rock shelters that were termed as such by the literary writers of old tourist guides. They also misguidedly called two of the tunnels caves; this however does not detract from the interest of from this unique folly garden that can be visited most times of the year.

Early articles on the caves were written by a Mr Lench in 1891 and shortly after by a Mr Davies in 1894. The most informative work is the *Hawkestone Handbook* of 1938 which included plans and locations of all of the caves and tunnels.

There are two features on Red Castle Hill. The first is a curved passage 12m long with a ceiling height of 1.8m, the second is a smaller passage just 5m long. It is more interesting because it is forged in a zigzag fashion so visitors fumble their way through.

There is only one subterranean feature on Elysian Hill: a manmade cave 1.5m wide by 2m long which was formerly used by the gardeners as a tool store.

The first cave visited on the terrace is found to the extreme south of the scarp near to the folly building and termed the White Tower. This is a naturally formed horseshoe shaped cave into which has been carved a rock cut bench. Travelling north along the escarpment is a cave known as Reynard's Banqueting House. It is aptly named from the numerous rabbit and poultry bones found in this natural rock shelter, that was described as a cave in the 1891 tourist guide. Following the path northwards for some distance we come across another portal which leads into St Francis' Cave. This is a slightly misleading statement in that the explorer is traversing through a long, snaking passage approximately 18m long, 1m wide and with an ample ceiling height of 2m. There are pick marks in profusion left from its excavation and the passageway at one time had a hinged door at its entrance.

The woodland footpath quickly takes us to another rock cut passage that is 8m long. A short flight of steps are found to its central portion and, as we leave the portal, we find nearby a rock alcove known as the retreat. This has a lofty ceiling height of 4m and a width of 3m. This particular feature is mentioned as a cave in the Victorian journals. A short walk and we find ourselves on Grotto Hill where the main underground features are found. We start with two tunnels, the first a short 5m arched heading that allows one of the garden paths to traverse a sandstone abutment that blocked its way. The second tunnel is more interesting in that there are two offset headings that meet into a circular chamber that hosts a central support pillar, a small window streams short rays of light to the central area so as we can dimly make a route through its extremities.

The final and most impressive feature that gives the hill its name is Hawkestone Grotto. It has been conjectured that the grotto originally started out as an early Roman metal mine, excavating copper deposited by thermal fluids into the sandstone strata. Historians in 1879 and 1892 forwarded theories, but unfortunately there is no real proof to add other than that Roman artefacts have been found at Hawkestone Park.

Hawkestone Grotto is a magnificent folly containing two circular chambers of 8m and 6m diameters, interlinked to these is a main chamber of random shape. This main chamber is approximately 28m long by 9m to 12m in depth, measured at various points around the grotto.

There are twenty-six roofing pillars left in for support, there are ten skylight portals of which only five now loose in daylight. The grotto contains four entrance passages, the longest and most northerly passage, 80m long, was added by Sir Rowland Hill, indicating that parts of the grotto pre-date the garden improvements. There are traces of copper minerals malachite and azurite in the walls of the main chamber, adding to the supposition that the grotto was formerly an old mine working. Another supporting factor is the the irregular-shaped main chamber. It is often conjectured that a hermit resided in the cave at some early date.

Sir Rowland built on this mythology by building a papier mâché automaton that resembled a hermit. This, however, was housed in one of the smaller caves. The grotto was visited by the famous Dr Johnston in 1774 and a short transcript was written into his diaries in which he pronounced it to be a place of terrific grandeur.

A description of the grotto exists from 1807:

The grotto is without anything of that diminutive or formal decoration and petitesse by which other grottoes are usually rendered.

The grotto was still very popular in 1891 and in pristine condition, as an article of the day states:

The visitor is introduced to the magnificent grotto, a vast subterranean cave, in the midst of which is a spacious recess fantastically inlaid with a great variety of shells, fossils and other curious petrifications.

It is said that the decorations were completed in 1790 by the two young Miss Hills. The great undertaking took them three years. A little of this decorative plasterwork remains to this day.

A number of windows are cut through the rock face that overlook the steep cliffs. They were originally fitted with stained glass, representing the four seasons and a philosopher at his studies.

The gardens and caves gradually became rundown during the Second World War. The park was being used at that time for military purposes, not least a prisoner of war camp. The caves and grotto began to be vandalised. There is a tale told of how, just after the war, two travelling cyclists tried to get refreshments at the nearby hall and, when refused, took out their tempers on the grotto, doing much damage in the process. During the 1970s the grotto became a favourite spot for local youths. An old works acquaintance told me how he would camp out in the grotto and have camp fires, and he often wondered how one of them had not fallen to an early death climbing in and around the grotto and cliff faces. The caves during this sedated period were host to small colonies of bats, but unfortunately now the park is open to continuous visiting and the bats have moved on. The good side to the park being reopened is that the grotto and caves are now maintained for posterity.

I recently found a small article concerning a man of troglodytic nature, a Richard Manford. He was presumed to have spent most of his life in a cave near Hawkestone. The unfortunate soul was found dying near to the main road on 10 May 1910. I am unclear as to whether the cave lay in the grounds of Hawkestone Park or whether it was a cave nearer to Market Drayton.

SELECTED BIBLIOGRAPHY

Middleton, Terry, 'Caves and Mines Hawkstone Park, Salop', *Cave Science*, (Vol.14, No.3, 1987)

Raven, Michael, *Gazeteer Shropshire*

Kempe, David, *Living Underground* (Herbet Press, 1988)

Bridgnorth (Notes by the late Mr E.H. Pee)

The Hawkstone Handbook, An Illustrated Guide (Wilding & Son Ltd, Shrewsbury, 1938)

Davies, R., *Hawkstone Park Illustrated Guide* (1894)

Jones, B., *Follies & Grottoes*, (1974)

Leach, F., *The County Seats of Shropshire* (1891)

Heathcote, John, *Cambridge Underground* (1980)

Shropshire Caving and Mining Club Journal (1980)

Baker, N., *Current Archeology* (September 1998)

Morris, John (Prof.) & Roberts, George, 'Carboniferous Limestones of Oreton/Farlow', from *Proceedings of the Geological Society* (1862)

Murchison, Roderick Impey, *The Silurian System* (1839)

Culingford, C. *British Caving* (1953)

Jackson, G. *Shropshire Folklore* (1883)

Timmins, H.T., *Nooks and Corners of Shropshire*

Byford Jones, W. *Shropshire Haunts of Mary Webb* (1937)

Cartlidge, J.C.G. Revd, *The Vale and Gates of USC-Con.* (1915)

Thackeray, V., *Tales From the Welsh Marches* (1992)

Kissack, P., *The River Severn* (1982)

Dyas, F.S., *Black Country Bugle* Article, *Bugle Annual* (1983)

Gandy, Ida, *Idler on the Shropshire Border* (1970)

Robinson, D.H., *The Wandering Worfe* (1980)

Wharton, Alan, 'The Underground Structures at Tong Castle', *Subterranea Britannica Bulletin*, No.10

Shropshire Mining Club Year Book (1961-1962)

Waite, V., *Shropshire Hill Country* (1970)

Bird, V., *Exploring the West Midlands*

Victoria County History of Shropshire

Shropshire Archaeology Society (Vol.21, 1898)

Cambden's Britannia (Vol.3, 1806)

Macefield, W.J., *Bridgnorth As It Was* (1978)

Clay, Rotha Mary, *Hermits & Anchorites of England* (1914)

Porter, Sheena, *Jacobs Ladder, Bridgnorth Town Guide* (1907)

Gwilt, C.F., *Hermitage Caves, Follies & Grottoes,*

Webb, Alan, Cave Beer (Article) *Tourist Guide to Bridgnorth* (Evans, Edkins & McMichael, 1875)

Shropshire Review

Shaw, Joan, *A Shropshire Scrapbook* (1990)

Adams, D., *Caves Between Llanymynach and The Dee* (Shropshire Caving Mining Club)

Mining in Shropshire (Shropshire Books, 1996)

Adams, D., *Llanymynach Ogof* (Edited by A. Pierce, Shropshire Mining & Caving Club)

ARTICLES

Price, George, *Progress at the Farlow-Oreton Cave, Clee Hills* (Birmingham Enterprise Transactions, 1970)

Remains of Early Hominoid Settlements in British Caves (Article by Sheffield University)

Smith, H. *Bridgnorth Hermitage* (Transactions Shropshire Archaeology Society Vol.1, 1877)

Travels in Tudor England – John Leland's Itinery by John Chandler

Cunnington, H.L., *Lavingstone's Hole*, 1934 (Bridgnorth Historical Society)

Baring-Gould, S., *Cliff Castles & Cave Dwellers of Europe* (1911)